A Century of Ideas

Fundamental Theories of Physics

An International Book Series on The Fundamental Theories of Physics:
Their Clarification, Development and Application

Series Editors:
GIANCARLO GHIRARDI, University of Trieste, Italy
VESSELIN PETKOV, Concordia University, Canada
TONY SUDBERY, University of York, UK
ALWYN VAN DER MERWE, University of Denver, CO, USA

Volume 149

A Century of Ideas

Perspectives from Leading Scientists of the 20th Century

edited by

B.G. Sidharth

B.M. Birla Science Centre,
Adarsh Nagar, Hyderabad, India

 Springer

Editor
Dr. B.G. Sidharth
B.M. Birla Science Centre
International Institute for Applicable Mathematics
  & Information Science
Adarshnagar
Hyderabad
iiamisbgs1@yahoo.com

ISBN 978-1-4020-4359-8   ISBN 978-1-4020-4360-4 (eBook)

Library of Congress Control Number: 2008923553

Dedicated to the memory of

Mr. Braj Mohan and Mrs. Rukmani Birla

Preface

The B.M. Birla Science Centre, inspired by the memory of Mr. and Mrs. B.M.
Birla, has blossomed into India's premiere institution for the dissemination of
science. Shortly after its inception, it started a series of lectures, most of them
aimed at a general audience or an audience of non-specialist scientists. Some
of the greatest minds in science have delivered these lectures over the years,
most of them under the B.M. Birla Memorial Lecture series and some under
the Distinguished Lecture series.

In fact twenty Nobel Laureates and a number of other scientists of equal
caliber have delivered these lectures. The distinguished list includes: Nobel
Laureates Professors William Fowler, S. Van Der Meer, Lord George Porter,
Antony Hewish, Norman Ramsey, Aarun Klug, Ilya Prigogine, De Gennes,
Werner Arber, Klaus Von Klitzing, Roald Hoffmann, Charles Townes, Ger-
ard't Hooft, S. Chu, Norman Borlaug, James Watson, The Dalai Lama, John
Kendrew, Prof. Herald Kroto and the eminent scientists include Fred Hoyle,
Sir Hermann Bondi, Prof. Jeffrey Sachs, Prof. Philip Morrison, Prof. Yuval
Ne'eman, Prof. Jogesh Pati, Prof. David Finkelstein, Prof. Walter Greiner and
so on.

Most of the lectures were on Physics and Astronomy, but some lectures
have been on Life Sciences and Chemical Sciences. For example those of Prof.
James Watson, Prof. Werner Arber, Prof. Norman Borlaug, Prof. Aarun Klug,
Sir Harry Kroto and others.

The present collection consists of most but not all lectures delivered in
the fields of Physics and Astronomy. The contributions of Prof. Abdus Salam,
Prof. Ramsey, Prof. Klitzing, and Prof. Steven Chu were unfortunately not
available to be included in this collection. Most of these lectures are accessible
to a wide range of non-specialist readers.

The lectures in this volume cover a very wide range of frontier topics,
starting with ideas of the Steady State cosmological model in Fred Hoyle's
article through the early formation of elements in nucleosynthesis in Prof.
William Fowler's article, through the long term picture of particle accelerators
in Prof. Van der Meer's article, the excitements of the discovery of Pulsars in

Prof. Antony Hewish's article, a critical discussion of irreversibility in Prof. Prigogine's article, a peep beyond the standard model in Prof. Ne'eman's article, a discussion of the standard model methods in Prof. 't Hooft's article, to topics like the apparently innocent soap bubbles in the article of Prof. De Gennes.

Wherever possible I have made an attempt to retain the original flavour of the lectures, which was a great part of the charm.

B.M. BIRLA SCIENCE CENTRE,   B.G. SIDHARTH
HYDERABAD, INDIA
JUNE 2005

Contents

Preface VII

List of Contributors XI

Fifty Years of Cosmology
Fred Hoyle 1

Science as an Adventure
Hermann Bondi 13

The Early Universe
William Fowler 19

The Long-Term Future of Particle Accelerators
Simon van der Meer 27

Energy and Evolution
George Porter 43

The Wonders of Pulsars
Antony Hewish 55

Is the Future Given? Changes in Our Description of Nature
Ilya Prigogine 65

Bubbles, Foams and Other Fragile Objects
P.G. de Gennes 77

**Beyond the Standard Model: Will it be the Theory
of Everything?**
Yuval Ne'eman 87

Living Joyfully with Complexity in Chemistry and Culture
Roald Hoffmann 101

A Confrontation with Infinity
Gerard 't Hooft 109

**The Creative and Unpredictable Interaction of Science
and Technology**
Charles Townes 123

**The Link Between Neutrino Masses and Proton Decay
in Supersymmetric Unification**
Jogesh C. Pati 139

The Nature of Discovery in Physics
Douglas D. Osheroff 175

**Symmetry in the Micro World – A Conversation with Nobel
Laureate Eugene Wigner**
B.G. Sidharth 205

List of Contributors

B.G. Sidharth
B.M. Birla Science Centre
Adarsh Nagar
Hyderabad 500 063, India

Fred Hoyle
University College
Cardiff
U.K.

Hermann Bondi
69 Mill Lane, GB – Nimpington
Cambs, Cambridge CB4 4X,
U.K.

William Fowler
California Institute of Technology
Pasadena
CA 91125, U.S.A.

van der Meer
CERN
Geneva
Switzerland

George Porter
Imperial College of Science
Technology & Medicine
Prince Consort Road
London SW7 2BB

Antony Hewish
Cavendish Laboratory
Madingley Road
Cambridge CB3 0HE, U.K.

Ilya Prigogine
Campus Plaine ULB
CP. 231, Bd. Du Triomphe
B-1050-Bruxelles, Belgium

P.G. de Gennes
College de France
11, place Marcelin Berthelot
75231 Paris Cedex 05

Yuval Ne'eman
Tel-Aviv University
Israel 69978

Roald Hoffmann
Cornell University Ithaca,
NY 14853-1301, U.S.A.

Gerard't Hooft
University of Utrecht
Utrecht 3584CC, The Netherlands

Charles Townes
University of CA Berkeley
366 LECONTE HALL
Berkeley, CA 94720, U.S.A.

Douglas D. Osheroff
Department of Physics
Stanford University
Stanford, CA 94305-4060, U.S.A.

Jogesh C. Pati
University of Maryland
College Park
MD 20742-4111, U.S.A.

Fifty Years of Cosmology

Fred Hoyle

Department of Applied Mathematics & Astronomy,
University College, Cardiff, U.K.

Fig. 1. Fred Hoyle delivering the B.M. Birla Memorial Lecture

Fred Hoyle was born on June 24, 1915 in Bingley, Yorkshire. His father, Benjamin was a wool merchant and mother Mabel was a teacher. Child is the father of man, it is said and certainly the young Fred was something of a rebel and even a truant. By age four he could reel off multiplication tables up to twelve and by age ten could navigate by the stars.

After early Grammar School, Hoyle studied mathematics at Emmanuel College, Cambridge, receiving his BA, winning in the process the Mayhew Prize in the Mathematical Tripos. After more distinctions and the Master's degree in Physics in 1939 Hoyle earned a fellowship at St. John's College, Cambridge. That was also the year he married Barbara. For six years during World War II he was developing radar technology with the British Admiralty. During these years he also started working with the well known Raymond Lyttleton on problems of accretion of dust and gas around large bodies. This was the precursor to his later interest in the origin of planetary systems and his belief that life must be a frequent occurrence in the universe.

B.G. Sidharth (ed.), *A Century of Ideas*, 1–12.
© Springer Science+Business Media B.V. 2008

In the early 1940s, Hoyle worked on his theory of stellar evolution, expanding on the work of Hans Bethe on the energy production within stars via a sequence of nuclear reactions. Hoyle was able to put forward a theory of nuclear fusion which could even account for the heavy element content in the solar system. This was an improvement of Bethe's work.

His work during the War years also brought him in touch with Hermann Bondi and Thomas Gold. He developed, as a result a continuous creation or steady state theory of the universe, though his perspective was rather different compared to that of Bondi and Gold.

Hoyle returned to Cambridge in 1945, after the war as a lecturer in Mathematics. In 1946 he authored two seminal papers, one on 'The Synthesis of Elements from Hydrogen' and the other on 'The Origin of Cosmic Rays'. In this latter paper he predicted that heavy elements would be found in cosmic rays, and this was confirmed after twenty two years.

In 1964 the accidental discovery of the Background Cosmic Microwaves by Penzias and Wilson confirmed the alternative theory of the universe, put forward most recently by George Gamow. This was a blow to Hoyle's steady state theory. Nevertheless, Hoyle continued to believe that the steady state theory with some modifications would be the ultimate theory. As late as the mid nineties he put forward along with his former student Jayant Narlikar, arguments in favor of steady state theory using ideas of electrodynamics.

Undoubtedly Hoyle's most important contribution to science was his work on the Origin of the Elements via Nucleogenesis inside stars. In this 1950s work Hoyle collaborated with William Fowler and the husband and wife pair, the Burbidges. This work, which has been described as "monumental" earned William Fowler the 1983 Nobel Prize – but Hoyle was excluded.

I asked Hermann Bondi about the possible reason for this exclusion. His answer was an endorsement of the view expressed by Nature. The Nobel Prize to Hoyle, in short, would also sanctify his other off beat ideas, for example his belief that life must be a frequent occurrence in the universe and that it had been transported to the earth, for example through bits of comets or meteorites. Hoyle and his collaborator Wickramasinghe had gone on to argue that certain epidemic causing viruses were also brought to the earth by the meteorites. While the Swedish Academy, as a sop later awarded Fred the prestigious Crafoord Prize in 1997, Fowler himself acknowledged his debt to Hoyle in his Autobiography written for the Nobel Foundation: "Fred Hoyle was the second great influence in my life. The grand concept of nucleosynthesis in stars was first definitely established by Hoyle in 1946."

Fred Hoyle held many prestigious positions and also received several honors. He was the prestigious Plumian Professor of Astronomy and Experimental Philosophy at Cambridge, as also the first Director of the University's Institute of Theoretical Astronomy. He held these positions till he resigned from them in 1972 and 1973. He was elected a Fellow of the Royal Society in London in 1957 and was knighted in 1972. He was also the Member of the Scientific Research Council from 1967 to 1972. He was the Chairman of the Anglo-Australian Telescope Board in 1973. He received several honorary doctorates, medals and prizes by learned societies and international organizations including the Royal Medal of the Royal Society, the Kalinga Prize of the United Nations, and the Balzan Prize. He died in Bournemouth in 2001.

Mention must be made of two other facets of Fred Hoyle. He was a prolific writer on popular science as also science fiction. His "A For Andromeda" was made into a television serial while his "Rockets in Ursa Major" was produced as a play.

The other aspect was his study of Stonehenge amidst claims that it was a primitive astronomical observatory. Fred's conclusions, were again, radical. Stonehenge indeed was an observatory of sorts.

At the age of 57, Fred Hoyle retired from his formal appointments in the UK, though he continued to hold honorary research professorships for example at the University of Manchester, at Cardiff, at Caltech, at Cornell and elsewhere.

A few years before Fred passed away he sent me a copy of his Autobiography "Home is Where The Wind Blows". In this he noted, "After a lifetime of crabwise thinking, I have gradually become aware of the towering intellectual structure of the world, one article of faith I have about it is that, whatever the end may be for each of us, it cannot be a bad one."

Indeed Fred had come a long way since the early years, in which he was distinctly aethiest. Later his vision developed into one contained in India's ancient metaphysics, that the universe is in some all permeating way intelligent.

I had the pleasure of hosting three lectures by Fred on different occasions. This prompted a peeved response from the head of the British Council who complained that for several years they had been trying to arrange a lecture by him, but without success. On one occasion, India's National Television was pestering him for an interview, to which he did not agree. At the same time an Educational Television outfit approached me to request him for an interview. When I told Fred Hoyle that this was a purely educational television channel, he readily agreed and spoke at length for about an hour. He made many interesting points during this interview, one of them being that the world is a troubled place because science has debunked religion on the one hand, and on the other hand has failed to provide an alternative.

On another occasion I encountered his radical thinking. There was an obscure hymn from India's Vedic literature, which if properly understood meant that the planets Mercury and Venus have phases. This was thousands of years before telescopes had been made. I asked Fred if this could be so. He thought awhile and said, "Yes I would say that this is possible. You do not need a telescope in the usual sense of the word. By placing your eye near the focal point of a highly polished spherical structure, even with a polished spherical bottom, you could see the phases quite clearly." His view was that intelligence is not a characteristic of modern man alone – even the builders of Stonehenge, for example, were very clever.

I graduated at Cambridge University in 1936, the year in which Hubble and Humason published their famous paper on the redshifts of galaxies. I had studied mathematics as an undergraduate, taking Part III of the Mathematical Tripos in my final year, my main topics in that examination being quantum mechanics, statistical mechanics and relativity, the latter both special and general. So from a student point of view I already knew a little about Cosmology at the time that Hubble and Humason published their paper. In particular, I knew about de Sitter's cosmological model which was to play an important role in subsequent years.

The worrying situation at that time in cosmology, as it seemed, turned out to be a relatively minor matter, namely the choice of suitable coordinates. Even the best-known cosmologists – de Sitter, Eddington and Lemaitre – had chosen coordinates appropriate to localities in the universe rather than the whole. This produced a sense of mystery that was more apparent than real as to what happened at the boundary of a locality. It is one of the features of Einstein's general relativity that when you choose coordinate systems with special properties you can mistakenly come to think of the properties as physical instead of as mathematical artefacts. Early workers on gravitational waves thought they were investigating physical waves when in fact the waves were in their coordinate system, and a similar situation existed in cosmology.

It was also in 1935–36 that this situation was put right, by H.P. Robertson in the United States and A.E. Walker in Britain and the resulting choice of coordinates later became known as the Robertson-Walker line element. Then in 1937 Robertson published an important article on cosmology in the Reviews of Modern Physics, which unfortunately I didn't read at that time because my research interests were in quantum mechanics and nuclear physics.

During the second world war it happened that Hermann Bondi and I worked closely together, and we continued to do so for a year to two after the war when we both returned to Cambridge as Junior Lecturers in Mathematics. My interests were now in astrophysics, and when Bondi decided he was going to make something of a speciality in general relativity, I joined him in that study. So it came about that at last in 1945–46 Bondi and I went in great detail through Robertson's article in the Reviews of Modern Physics, looking carefully into its fine points as well as into the broader arguments.

The cosmological models favored by Robertson were of the so-called Friedmann type, which is to say what today we would call big-bang models, the idea being that the universe originated suddenly all in a moment. This view had a fairly wide constituency at that time in the United States, due in a considerable measure to Robertson himself and also to George Gamow, but not in Europe. Many Europeans felt the theoretical conclusion of a big-bang origin, arrived at in Robsertson's analysis, was a product of simplifying assumptions in the analysis. Notably, it was felt that the assumptions of isotropy and homogeneity were constraining influences on the problem. Lifshitz in the USSR, the collaborator of Landau in the famous Landau-and-Lifshitz textbooks, published an extremely long and complicated paper, in which he claimed that because of departures from homogeneity, the big bang was an invalid concept. And at every conference on cosmology and relativity held in Europe in the 1950s, Otto Heckmann, the Director of the Hamburg Observatory made a similar claim, at first with respect to departures from isotropy and then with respect to inhomogeneities. I personally found these claims both disturbing and irritating, because coming from people of high standing I felt I ought to understand them and I couldn't. Heckmann in particular was always telling me that some especially clever student of his had demonstrated the matter beyond dispute. Eventually in the early 1960s I had an especially clever student of my own in the person of Professor Jayant Narlikar, now of the Tata Institute of Fundamental Research. The first issues I asked Narlikar to investigate were the claims of Lifshitz and Heckmann. Rather as I had expected he found them to be wrong. Narlikar was soon able to offer a simple proof that departures from homogeneity and isotropy cannot in themselves prevent the phenomenon of the big-bang. A more ambitious proof of the same result was given a few years later by Hawking and Ellis. To complete this aspect of my story, it must have been in 1970 or thereabouts that Narlikar and I published a very different idea for invalidating the big-bang, namely that the effects of quantum mechanics would need to be considered at the earliest moments of the universe, and as such might make the concept of the big-bang meaningless. Since then, Professor Narlikar and his students have proved this to be the case. Because of quantum mechanics there can be no big-bang in the sense the concept is widely used by those many commentators in the media and even in scientific journals, commentators who are unfortunately all too often ignorant of quantum mechanics, or at least of its subtler aspects. To put the matter a little more technically, because of quantum mechanical uncertainties in the line element, space time singularities do not occur, a result that is also applicable to so-called black holes, vitiating many of the things which are commonly said about black holes.

Meanwhile as early as 1947–48 a few of us in Cambridge were investigating a new physical idea in cosmology, namely that matter might be subject to a continuous form of creation. At first, Hermann Bondi would have none of it, although his close friend Tommy Gold was rather in favor of it. I was myself neutral to the idea. I realized in 1947, when Bondi and Gold turned to other ideas, that if continuous creation were to have any hope of acceptance it would have to be given a mathematical expression. In the latter part of 1947, I came to the conclusion that a new form of field would be needed, and that a scalar field was not only the simplest possibility but also the most promising. I wrote the field on paper as a Capital C, and from then on it became known as the C-field. In January 1948 I found how to use the C-field in a modification of Einstein's equations with the result that the equations had as a particular solution what became later known as the Steady-State model. This, let me emphasize, was not a static model but one in which the main features of the universe are steady like a steadily flowing river. The universe expands but it does not become increasingly empty because new matter is constantly being created to make up the deficit produced by the expansion. By the end of February 1948 I had written my paper, "A New Model for the Expanding Universe", in the form in which it was eventually published in the Monthly Notices of the Royal Astronomical Society, after being rejected by the Proceedings of the Physical Society and by the Physical Reviews. On 1 March 1948 I gave a colloquium on the new model at the Cavendish Laboratory, at which both of the two great pioneers of quantum mechanics were present, Paul Dirac and Werner Heisenberg. Heisenberg had been invited to Cambridge for a six-months period, which he had spent at my own college, St. John's, and because of this I had got to know him quite well. Possibly this was the reason why I heard later that, after his return to Germany, he had said that the concept of a steady-state model was the most interesting thing he had heard during his stay in Britain.

Naturally I showed my paper to Bondi and Gold: Bondi saw immediately that his difficulty about the conservation of energy has been answered. The remarkable thing about the C-field was that its energy density was negative. As matter with positive energy was created, the energy of the C-field became more negative. In the flat space time of special relativity this would have led to a creation catastrophe, with matter being created at an ever increasing rate as the C-field became more and more negative. But in general relativity, which is to say with gravitation present, this did not happen. The C-field use gravitationally self-repellent, so that as matter was created it was forced apart by the C-field, thereby maintaining a steady balance. Thus at a stroke two crucial features of the universe were explained, its matter content and its expansion. Neither had to be arbitrarily assumed, as in the big-bang models which had been discussed by H.P. Robertson. The universe expanded because it was forced to expand, not because it had arbitrarily been created in a state of explosion.

In March and April of 1948, Bondi and Gold then conceived of a remarkable point of view. Instead of regarding a steady-state universe as a deductive consequence of a set of mathematical equations, as one normally does in theoretical physics, they conceived of it as a philosophical axiom, which they referred to as the 'perfect cosmological principle'. From their perfect cosmological principle they were then able to obtain the same geometrical structure for the universe as I had obtained from the mathematical equations, namely what is usually known as the de Sitter line element, and thereafter the discussion became similar in its astrophysical consequences to mine. By about May I think it was, they had written a paper to this effect which they sent to the Monthly Notices of the Royal Astronomical Society. In the ensuing months there occurred the first of the circumstances which have sometimes caused me to regret that I ever had anything to do with cosmology, for owing to my paper being rejected by two journals, the time delay involved in its successive rejections led to its eventually being printed several weeks later then the Bondi-Gold paper. The lesser aspect of this inversion in printing of the order in which the papers had actually been written led to a mix-up over priority, which was compounded by the fact that Bondi and Gold had actually discussed the steady-state idea verbally, but without doing anything definite about it, as early as 1946, ahead of my C-field idea of 1947. So in the general confusion I was never able, even to myself, to make up my mind as to exactly how the history should be fairly stated, and in the event I decided to say nothing at the time, leaving the situation to come out as it would.

Of much more scientific relevance than priority, there was an immense difference of emphasis between my paper and the Bondi – Gold paper. My paper simply said: "Here is a new cosmological model to be discussed along with other models". Because of their central philosophical axiom, Bondi and Gold could not take this guarded position. They had to come out and say assertively that of necessity the steady-state model must be the correct model. No question about it from their point. This had two main effects: It caused the theory to be attacked more ferociously than my point of view would have done, and it provoked far more discussions than mine would have done.

Although throughout the 1950s the three of us were thrown together in order to defend the theory, I must confess that I did not myself have much liking for the physical aspects of the 'perfect cosmological principle' and already in 1949 I wrote a paper critical of it. The perfect cosmological principle required the universe to be unchanging with respect to time. But quite evidently localities within the universe are indeed changing with time. What was it that decides, I asked, the scale at which there is a change from local change to universal invariances? This question has lain for almost forty years unanswered. As I shall indicate at the end of my talk, it may well be the most relevant question of all, unfortunately asked long before its time. My suggestion in 1949 as to its answer would have been that localities deviating from a steady-state condition might have dimensions of the order of thirty million light years. By the early 1960s, Narlikar and I had increased this estimate ten fold, to about three hundred million light years. Today I would increase it to at least the greatest distances at which galaxies and quasars are observed.

The 1950s were noteworthy for two quite different developments, one theoretical, the other observational. The theoretical challenge was to improve the mathematical elegance of the theory. I made the mistake myself of sticking to the physical equations, whereas a friend, M.H.L. Pryce, used a more abstract approach known as an action principle. This gave a better classical formulation of the steady-state theory.

On the observational side, attempts were made to disprove the steady-state theory by showing that the astrophysical properties of galaxies change with time, which is to say with respect to red-shift. Many disproofs were claimed. Some were withdrawn and others were maintained with increasing emphasis as the years passed by. Tommy Gold was the most outspoken of us in reply to these criticisms. He pointed out that the greater the distance of a galaxy the fainter it became and the more difficult it necessarily was to make observations accurately. Tommy's point was that progressively increasing errors with increasing distances were being falsely interpreted as physical changes–in other words the claims were artefacts arising from errors of observation. In all cases known to me, this riposte from Gold has turned out to be correct. With the much more sensitive observations available today, no astrophysical property shows evidence of evolution such as was claimed in the 1950s, to disprove the steady-state theory. In particular, the strong claims of Martin Ryle have turned out to be wrong, as Gold always said they would be. Technically speaking again, the luminosity function of radio galaxies has turned out to be invariant with respect to red shift, the opposite of what Ryle claimed. If all this had been known in 1960, the steady-state theory would then have been considered proven, and the development of cosmology following the discovery of the microwave background in 1965 would probably have been very different.

In 1963–64 I gave a course of lectures at Cambridge on relativity and cosmology in the preparation of which I went carefully over the work of George Gamow and his colleagues on the synthesis of light elements in a hog big-bang model of the universe. It seemed that a calculation of the helium/hydrogen ratio to be expected in such a model could be improved, and together with Roger Taylor I set out to make the necessary calculations. In such a model there is a present day microwave background temperature. Taylor and I found that if we knew this temperature we could calculate a cosmic value for the hydrogen/helium ratio, and vice versa, if we knew a cosmic value for the hydrogen/helium ratio we could infer what the present day microwave background temperature had to be. But when we examined the astronomical literature concerning hydrogen/helium ratios determined by astronomical means, we found a wide range of values corresponding to helium abundances by mass, ranging from a low of about 15% to a high of about 40%. This was vastly too broad a range for anything useful to be inferred about a possible microwave background temperature.

It must have been in 1964 that I was sitting beside Lake Como in Italy, with Bob Dicke from Princeton University Dicke told me that his group at Princeton were setting up an experiment to look for a possible microwave background, and that they were expecting a temperature of about 20 K. I said this was much too high, because a background – if there was one – could not have a temperature above 3 K, the excitation temperature of molecular lines of CH and CN found by Mckellar in 1940. Shortly after that the background was found at the Bell Telephone Laboratories by Penzias and Wilson, and it had a temperature almost exactly on Mckellar's value. The big mistake Bob Dicke and I had made was not to realise we had it there beside Lake Como, in our coffee cups. However carefully one guards against it, opportunities like this come and then slip away through one's fingers.

The discovery of an actual microwave background made it profitable to calculate the light element synthesis problem more ambitiously than Taylor and I had done, and in 1966–67 Bob Wagoner, Willy Fowler and I set ourselves to do this. Interesting results were confined to just four light nuclei, D, ^3He, ^4He, and ^7Li. From the results we were able to show how astrophysical measurements of cosmic values for these light nuclei could be used to infer the properties of a hot big-bang model, assuming the latter to be correct and assuming the astrophysical measurements to be truly cosmic. Since then, Bob Wagoner has periodically updated these calculations, obtaining slightly differing results as the physics of the problem has changed somewhat over the years, for example by there being three types of neutrinos instead of the two types used in the first calculations.

This work with Wagoner and Fowler was my last essay in cosmology by orthodox methods. For the following eight years, up to 1975, I was heavily occupied in administration, and since 1975, my thoughts have run in other directions. This does not mean that I have lost interest in cosmology, but rather that I have sat around waiting for something significantly new to happen. Despite immense numbers of people swarming into Cosmology in the United States, in Europe and in the Soviet Union, nothing very profound seems to me to have happened over the past twenty years. Ask one of the younger generation what evidence they would offer for the correctness of a hot big-bang model and the chances are that they would say, first, the existence of the microwave background, and, second, the synthesis of D, ^3He, ^4He, and ^7Li. After that there wouldn't be much to offer. So the situation remains essentially as it was in 1965–66, which I regard as a distinctly bad omen for the theory. Always in the past, whenever a correct theory has been established, a decade or more of rapid progress has been forthcoming, not a state of stagnation, more or less. Indeed the one interesting thing to emerge in the early 1980s was a partial reversion to the steady-state model, which came about in the following way.

From a properly based scientific point of view the discovery of the microwave background did not come as an unmixed blessing to protagonists of the hot big-bang model. It was soon found that the background had a remarkable large-scale isotropy. It was almost the same coming from regions of the sky diametrically opposed to each other, despite such regions never having been in communication with each other at any time in the past in such a model. The attempts made to explain this large-scale model is to fall back on the arbitrary supposition that the background was isotropic because it was created that way at the origin of the universe. Indeed the model required every important observable aspect of the universe to be derived from the manner of its creation, essentially making it just as impossible to understand anything on rational scientific grounds, as if one were to believe in the first page of the Christian Bible, which actually, I had to suspect was playing an important role in the minds of those who supported the model.

Narlikar and I, already in the early 1960s, had given an explanation of isotropy and homogeneity in terms of the steady-state model, which after a sufficient number of generations simply expands away initial irregularities, just as irregular motions in a gas disappear if the gas expands adiabatically to a sufficient extent. From about 1980 onwards this idea was taken over in what were called inflationary scenarios. An inflationary scenario has three parts to it. There is an initial big-bang, then a steady-state phase, even in some scenarios, down to the operation of a similar scalar field to that which I had postulated at the end of 1947, and finally there is a freely expanding phase like one of the Friedmann models studied by H.P. Robertson. The initial big-bang has no observable function in such a scenario, for nothing in it lives through the steady-state phase to come down to us today as an observable entity. The microwave background, the creation of matter, and the irregularities which become the galaxies, all belong to the second phase, the steady-state phase. Thus the initial big-bang is superfluous like the attempts made in the 1930s to give quantum mechanical system unobservable internal variables in the hope of restoring determinism. An inflationary scenario functions just as well if the initial big-bang is omitted, in which case the universe is steady-state followed by a freely expanding Friedmann model, and is the same so far as astrophysics is concerned as a model studied by Narlikar and myself in 1966, a model which we referred to as a 'bubble universe'.

Our perception differed from an inflationary scenario only, so far as I can see, in them being dominantly mathematical rather than physical. The equations which relate the geometrical behavior of the universe to its physical content and to the creation of matter are non-linear, and it is a mathematical feature of non-linear equations that as well as possessing non-unique ordinary solutions they can possess a unique singular solution. What we found was that our equation had a unique singular solution and it was this solution that yielded the steady-state model with creation of matter. The ordinary solutions on the other hand were analogous to models of the Friedmann type without creation of matter. Our idea in the bubble universe was that there might be switches in particular localities out of a universal singular solution into a regional ordinary solution, yielding a Friedmann type bubbles embedded in a steady-state universal ocean. Or of course many such bubbles. The difference between this model and an inflationary scenario is that our switches in the mathematical solutions were conjectural, whereas it is now claimed in an inflationary scenario that the switches can be understood in terms of modern supersymmetry theories in Particle Physics. If people generally and cosmologists in particular would get away from this fixation with a mock-biblical big-bang and think more about the relation of a bubble universe with Particle Physics in it, then I think there would be the best chance of relating cosmology to astrophysics, a relation which to this point has been almost non-existent.

The first big issue to get straight is that the mean density existing in the basic steady-state cannot be superdense. Superdense conditions with supersymmetries in Particle Physics playing a dominant role, exist in local objects, not smoothly everywhere. The typical mean density is that which we observe in galaxies. Indeed the galaxies are aggregates of material left over in our particular bubble from its former steady-state condition, a mean density typically in the range 10^{-24} to $10^{-21} gm/cm^3$. Stars condense everywhere in the steady-state, producing a radiation background with a temperature that can be calculated to be necessarily around 300 K. In the expansion of our bubble the mean density has fallen by about a million and the temperature of the background by about a hundred. The origin of the radiation background was simply stellar radiation thermalised by immense quantities of dust.

This picture is close to being proven. Indeed I would say that it is proven, by the observed fine-scale isotropy of the microwave background. If the background came from a universal superdense state, where it had uniformity and isotropy, the later propagation of the radiation in non-uniform gravitational fields, which must have happened in such a theory when, clusters of galaxies form, should have produced measurable non-uniformity in the background on the scale of clusters of galaxies. This is not observed. Production of radiation by stars rather uniformly distributed, and by thermalisation due to dust, must produce an exceedingly uniform background on the other hand. This is because the dust, being rather insubstantial in its mass, would be subject to immense forces due to radiation pressure, were there any appreciable fluctuations in the untensity of the radiation background, causing the dust to adjust itself, quickly so as to produce a very unform situation.

The conventional notion that life originates on the Earth is the greatest running farce in scientific history. Anyone with a little physical sense should be able to see from the complexity and function of proteins that not even a single enzyme could ever be produced by random processes here on the Earth. Only if the whole universe is typically at a steady-state temperature of $300\ K$, over a very long span of time can the origin of life be understood in rational terms. The ideas currently held by biologists appeal just as much to irrational miracles as did the creationists who preceded them.

In the old steady-state theory of the 1950s and 1960s the balance between creation of matter and the expansion of the universe was thought to be stable. This I now think was a mistake, caused by regarding the creation process as being spatially uniform. If, however, creation occurs in localized objects, of which the quasars are perhaps the still recognizable remnants in our region of the universe, then quite likely the balance between creation and expansion is unstable, in which case every locality in the universe would be oscillatory, approximated to by a closed Friedmann model of finite volume except near minimum phase when the ambient C-field becomes strong enough to produce an intense burst of creation at a multitude of quasar-like centres of activity. The resulting sharp increase of the C-field then blows the locality back into an expanding phase. On this view the important question becomes to decide how big are the oscillating localities, thereby determining the period of the oscillations. On this view our locality, presently in an expanding phase will eventually fall back on itself, contracting until the mean density rises to 10^{-24} to $10^{-21} gm/cm^3$, the temperature of the microwave background rises to a $300\ K$, and the C-field becomes sufficient to produce another intense round of creation and of the birth of a multitude of stars, preceding yet another expanding phase. The picture is of a multitude of expanding and contracting bubbles with an immense flash of creation occurring as each bubble reverses from contraction to expansion, and with the whole ensemble of bubbles forming a kind of dynamic steady-state universe.

There is a further line of argument to which I attach great weight, that demands a universe of this type, and which rules out the purely Friedmann models. Many of our basic physical equations are time symmetric, as for instance the equations which determine the generation and propagation of light are time symmetric. The usual assumption that only the past-to-future solutions of such equations exist in the universe seems artificial and unsatisfactory. A theory in which both past-to-future and future-to-past solutions are generated equally in every local radiation process seems dictated by considerations of completeness and elegance. In such a time-symmetric theory what we observe locally is a sum of radiation generated locally, of radiation received from the past and of radiation received from the future. The latter is an addition to usual considerations, and by including it, the possibility exists that the sum of the three contributions for a time-symmetric theory turns out to be the same as for the usual time-asymmetric theory. For this to be possible, the universe itself must have an overall expansion which is of the steady-state type. The Friedmann models will not do, they give a wrong summation. This matter was first discussed some forty years ago, so far as classical theory is concerned, by Wheeler and Feynman. Narlikar and I showed in 1968 that the same result holds in non-relativistic quantum mechanics, and in 1970 we extended our proof to relativistic quantum electro dynamics. The fact that an overall steady-state structure for the universe permits local radiation processes to be time-symmetric and yet leads to the imposing of normal cause and effect on the flow of events has always been to my mind a guarantee that this form of cosmological theory will turn out to be basically correct.

Science as an Adventure

Hermann Bondi

Cambridgeshire, U.K.

Fig. 1. Hermann Bondi delivering the B.M. Birla Science Centre Distinguished Lecture

Hermann Bondi was born in 1919 to Samuel and Helene Bondi of Austria. He had his education at the Real gymnasium in Vienna. He went on to study at Trinity College, Cambridge. But during World War II he was interned as alien enemy. This gave him the opportunity to work with Thomas Gold and Fred Hoyle on the radar. He became a Fellow of Trinity College in 1943 and in 1947 he married Christine Stockman.

Around 1948 Bondi and Gold and from an independent perspective, Hoyle propounded the Steady State Theory of the universe. While the approach of Bondi and Gold was through a perfect cosmological principle, Fred Hoyle was working with the idea of a field out of which matter would emerge.

Bondi held various positions during his long and interesting career, starting as a temporary Experimental Officer for the Admiralty in 1942 through Assistant Lecturer in Mathematics in Cambridge University in 1945 and Lecturer in

B.G. Sidharth (ed.), *A Century of Ideas*, 13–18.
© Springer Science+Business Media B.V. 2008

1948. He was Professor of Mathematics at King's College, London in 1954 and became the Director General of the European Space Research Organization during the period 1967 to 1971. From 1971 to 1977 he was Chief Scientific Advisor to the Ministry of Defence in Britain, then Chief Scientist for the Department of Energy between 1977 to 1980 and Chief Executive of NERC between 1980 to 1984.

This apart he had become a Fellow of the Royal Society in 1959 as also Fellow of the Royal Astronomical Society. He was knighted in 1973.

He received numerous medals and honours and visiting professorships. In 2002 he was given the gold medal of the Royal Astronomical Society. He died in 2005. Bondi has a large number of publications – papers and books to his credit.

He visited the B.M. Birla Science Centre twice and delivered lectures under the Distinguished Lecture series. He spoke about Energy and also the Adventure of Science. The point that Bondi made was that advances in telecommunication would cut costs as well as dependence on fuels. For example one could have teleconferences or work from home on a computer – all this at a fraction of the transport and travel costs. Perhaps this is the shape of things to come. He was a rationalist, which means he was an aethiest – but he was also a deep humanist and in fact won the prestigious G.D. Birla International Award for Humanism in 1990. Above all he was a very delightful and thought provoking conversationist.

Many of our fellow citizens have an image of science and of scientists we would find hard to recognize: They tend to think of science as something rigid, firm and soulless (and generally dull) created in an objective and often solitary manner by cool passionless persons. While they might be willing to see some nobility in 'pure' science, this reluctant generosity does usually not extend to its 'dirty' offshoot, technology. Absurd as all this picture looks to us, it is, I think, worth examining how these dangerous misconceptions arise and what might be done to improve the understanding of what we do.

These views are indeed dangerous in several respects. First there is the angry puzzlement that arises when no firm clear answer can be given to scientific queries that are of public interest (usually in the environmental or medical fields). When on some such issue different scientists hold differing views, journalists speak of "this extraordinary scientific controversy". When I tell them that controversy is normal in science and is indeed the lifeblood of scientific advance, they find it hard to believe me. The public tend to think that at least some of the scientists involved in such a controversy must be either venal or incompetent or both. These views arise partly from the conflict between the popular view that scientists are 'objective and dispassionate' and the normality of active arguments, yet people are unwilling to abandon their view. Moreover, the piece of science most people are familiar with is the Newtonian description of the solar system (Newton's clockwork, as I like to call it). The rigid predictability of this is taken as a model of what all science should be like. When this expectation is not fulfilled, there is disappointment.

A model, familiar to all, for many fields of science is weather forecasting, but this is not appreciated. The curse of rigidity is thought to apply to us and this view is reinforced by teaching mainly the examinable pieces of science where the right/wrong classification can readily be applied.

Secondly there is the denigration of technology arising from the widely held view that it is purely derivative and trails science. Thirdly and in some respects most importantly, there is the worry that it is on the basis of these widespread misconceptions that young people make their career choices of whether to become scientists or not. I feel sure that some who would have made excellent scientists were frightened off by the rigid image ("to every question there is just one right answer") so often conveyed at school, while some of those who become scientists guided by this image are disappointed to find their work full of uncertainties and question marks. We need the adventurous souls, but do little to attract them. I sometimes comment that if we were a business with a prospectus as misleading as the one of science so often presented at school, we would be in trouble with the law.

How have these misconceptions arisen and what can we do to avoid generating them? I think there are a number of reasons. Foremost perhaps is the understandable desire to get the maximum quantity of science taught in the necessarily limited amount of time available (at school or even in undergraduate courses) with no attention paid to the need to convey something of its spirit. Coupled with this is the wish to confine instruction to the supposedly certain parts of science to avoid teaching something that later turns out to be incorrect. In fact it would be most educational to convey wonder and uncertainty. Neither the philosophy of science nor its history are considered to be parts of the normal syllabus. Yet it would be very beneficial to go through some of the very intelligent ideas of our predecessors that turned out to be wrong.

I totally accept that teaching hours are limited. If time is given to describing the evolution of scientific ideas and how they were shaped by technological developments, then clearly the total amount of science covered will be less than it is now. In my view this would be a price well worth paying for educational as well as for scientific reasons and would demonstrate the intensely human nature of science.

Perhaps an example will help. Children are taught that the Earth goes round the Sun, but rarely about tests of this hypothesis and probably never about the historical development which in fact is fascinating and, as will be seen, could readily be taught.

By the late seventeenth century the Copernican system was accepted by virtually all astronomers. The great prize and test would be measure a stellar parallax, the apparent change in the position of a near star due the Earth changing its position during its orbit about the Sun. The inaccuracies of the instrumentation of the time, coupled with the difficulty of choosing a sufficiently near star, led to numerous unsubstantiated claims until the first unambiguous parallax was at last established by F.W. Bessel in 1838. But already in 1725 James Bradley had discovered stellar aberration, the apparent change in the position of stars due to the change in the Earth's velocity during its orbit. This was the first clear test of the heliocentric system and should surely be widely taught at school (the effect on the inferred direction of a star due to the motion of the telescope is readily described). This would be far more helpful than the mere assertion that the Earth orbits the Sun. The aberration angle is many times greater than the parallax angle of the nearest stars, which accounts for it having been discovered much earlier. It is amusing to speculate that, had our civilization developed on Jupiter, with its bigger orbit, but lower velocity, parallax would have been much larger and aberration rather less than here, so that presumably parallax would have been found first.

This story is a good example of scientific evolution, showing how it is driven by technological advances (the gradual improvement in the precision of astronomical measurements which eventually enabled Bessel to measure so small a parallax angle successfully), but also how a good scientist works: Bradley had originally not thought of the then unexpected phenomenon of stellar aberration, but very speedily worked it out to account for his otherwise inexplicable measurement of a stellar position shift at right angles to the parallax shift he was expecting to find.

In the philosophy of science I am a follower of Karl Popper. He sees the task of a scientist first to propose a theory that of course needs to be compatible with the empirical knowledge of the day, but that also must forecast what further, future experiments or observations will show. If such are performed and are incompatible with the theory, we say that it has been disproved. Liability to empirical disproof is the defining characteristic of science. If the tests turn out to be consonant with the forecasts of the theory, we must never regard this as a proof of it, since it remains scientific only if it continues to be liable to be disproved by further experiments. Thus all scientific insights should be viewed as provisional only. It is because the wholly unexpected can happen that science is such an adventure.

This analysis is appropriate because theories make general statements, whereas experiments and observations inevitably deal with the particular. This is also the reason why a theory can never be deduced from empirical knowledge. It necessarily requires a leap of the imagination to formulate one. Equally it is imagination that is needed to devise a novel experiment to test a theory. Thus imagination is essential in science, but do our fellow citizens appreciate this? It is natural that a scientist will argue fiercely to defend a favored theory, perhaps by criticizing the accuracy or reliability of an incompatible experiment, which will be defended by its originator with equal passion. If relating this were part of ordinary teaching, perhaps the absurd popular picture of the cold, unimaginative, passionless scientist would gradually fade away. But there is another point on which we ourselves may be somewhat vulnerable. Communication is essential in science. We fully accept this. This need makes me think of a customer coming into a pet shop asking to buy a parrot of especially high intelligence. After some consideration he is sold a particular bird. Two weeks later this customer returns to the shop absolutely furious because this reputedly so intelligent bird has not said a word in all this time. However the pet shop owner replies: This parrot is a thinker, not a talker. Indeed we do not regard any work as part of science until it has been widely communicated through being published in the accessible scientific literature. Yet do we consider the teaching of communication skills to have a legitimate claim on the time table of a science course? We all have the experience of a graduate student, highly competent in the relevant topic, yet finding it immensely difficult to convey the results in understandable form by the spoken or written word. Most of us eventually learn communication skills on the job, but do we give their early systematic acquisition the priority it deserves?

Nor do we often analyze the means of the conveying of information in depth. To me the printed word is of only modest effectiveness, though its permanence and wide distribution make it essential. The formal lecture is rather more efficient, but the less formal seminar or workshop are far better for exchanging information. Yet to chat to a few colleagues with a glass in one's hand is superior to all other methods, for then one is willing to talk about one's doubts and failures as much as about one's successes.

I want to return now to the relation between science and technology, which is so often misunderstood. It is implicit in Popper's definition of science that tomorrow we can test our theories more searchingly and thoroughly than we can today. This means that the progress of science depends on the advance of the methods of empirical testing, i.e. of the available technology. Of course equally technology can get ahead by using novel insights of science. Thus it is a mutual relation in which neither science nor technology can be called primary, with the other secondary. Perhaps I can illustrate my thinking with an example of this interaction.

Physics made enormous strides during the last quarter of the nineteenth century with the discovery of the electron, of ions, of radio-activity, etc. Why were such discoveries made then and not earlier or later? Most of such work requires the use of evacuated vessels. Their availability depends on the efficiency and reliability of the pumps needed to extract the air. It so happened that the machining of brass pistons and cylinders improved considerably in the 1850s and 1860s. Though this was an essential pre-condition, it was not sufficient. Any vacuum system inevitably develops leaks that have to be plugged. For non-moving parts, sealing wax is an old established efficient means, but it is rather rigid and thus cannot be used in the links between the vibrating pump and the experimental vessel. A reasonably suitable material became available at the time, namely plasticine. (One could therefore say that much of physics is ultimately based on plasticine). The availability of such a vacuum was a major technological step that allowed much scientific work to be done.

In due course the use of such reliably evacuated vessels permitted Roentgen to make his great scientific discovery of X-rays a hundred years ago. Their importance for medicine was soon appreciated (though it took much longer before their dangers were understood). Accordingly, a new technology of X-ray machines came into being that in due course made them affordable, reliable, precise and safe. Some fifty years after Roentgen, these machines were used to study the structure of organic materials and thus the new science of molecular biology came into being. This in turn gave birth, in time, to a wholly new technology, bio-technology. This is a clear example in which each advance of science or technology leads to an advance in the other. Neither can claim primacy.

The international nature of science is so strong and pervasive, because science is well tailored to our universal human characteristics, above all to our fallibility. Similarly it suits our sociability and our need to communicate. We value imagination and ingenuity highly, but the supreme yardstick of empirical test is recognized by all.

I would like to conclude with a personal story which illustrates some of these features. Many years ago my late colleague R.A. Lyttleton and I investigated the consequences that would arise if the electric charges of the electron and the proton were not exactly equal and opposite. (At the time this was only known to one part in 10^{13}). We showed that there would be very interesting astronomical and cosmological consequences if the discrepancy were as small as one part in 10^{19}. This paper irritated many. In their desire to prove us wrong, several very ingenious experiments were devised which showed that the maximum permissible discrepancy was less than one part in 10^{22}, far too small for the effects we had calculated. So within a very few years we had been disproved. However, I am proud of this paper and in no way ashamed. Thanks to the work which it provoked, an important constant of nature is known to much higher accuracy than before.

Following Popper, we know that empirical disproof is the seminal event to science. One can be right only for a limited time, but to be original and stimulating is the essential contribution a scientist can make to the unending adventure that is science.

The Early Universe

William Fowler

California Institute of Technology, Pasadena, U.S.A

Fig. 1. William Fowler with Mr. G.P. Birla to his immediate left and Prof. J.V. Narlikar to his right

William Alfred Fowler was born in 1911 at Pittsburgh, Pennsylvania. He was raised in Lima, Ohio, from the age of two, as his parents shifted to this place. He had a great fascination for Steel Locomotives because of the Pennsylvania Railway Yard. In fact in 1973 he travelled on the Trans Siberian Railway from Khabarovsk to Moscow as a steam engine powered the train for nearly two thousand five hundred kilometers.

During his school days he was an accomplished football and baseball player. After school Fowler joined the Ohio State University, Columbus in Ohio to study ceramic engineering. However he soon became fascinated with Physics and transferred himself to the Engineering Physics department. Here he had to do all kinds of jobs for survival – as a waiter, as a dish washer, selling ham and cheese at the central market in Columbus and so on, earning five dollars for all his efforts. His undergraduate thesis was on "Focussing of Electron Beams", experimental work carried out under Prof. Willard Bennett.

B.G. Sidharth (ed.), *A Century of Ideas*, 19–25.
© Springer Science+Business Media B.V. 2008

On graduation Fowler joined the world famous California Institute of Technology as a graduate student for work under the famous C.C. Lauritsen in the Kellogg Radiation Laboratory. Fowler received his PhD in Physics in 1936 for work which showed the symmetry of nuclear forces between protons and neutrons. Thereafter he became an Assistant Professor at Caltech. However due to the second world war, the Kellogg Laboratory was engaged in defence research.

Lauritsen and Fowler reconverted Kellogg as a Nuclear Laboratory after the war and concentrated on nuclear reactions in stellar interiors. This was the starting point of nuclear astrophysics. Soon they confirmed that there was no stable nucleus at mass 8.

In 1951 E. Salpeter of Cornell came to Kellogg and showed that the fusion of three helium nuclei of mass 4 into the carbon nuclei of mass 12 could occur in red giant stars, but not in the big bang. Then in 1953 Fred Hoyle got an experiment to be performed in Kellogg which quantitatively confirmed the fusion process in red giants.

Hoyle had a great influence on Fowler. The original idea for stellar nucleosynthesis was first established by Hoyle in 1946 itself. Fowler spent a year in Cambridge, England in order to work with Hoyle, where they were joined by Geoffrey and Margaret Burbidge. The next year Hoyle and the Burbidges went to Kellogg and thus in 1957 they came out with a paper, "Synthesis of the Elements in Stars". This important work demonstrated that all the elements from carbon to uranium could be produced inside the stars, starting with the hydrogen and the helium produced in the big bang. William Fowler was awarded the 1983 Nobel Prize for his researches, along with S. Chandrasekhar.

Through all these years Prof. Fowler retained a sense of liveliness, cheer and humour. He would recount an encounter with the late Indian Prime Minister Mrs. Indira Gandhi, who attended one of his lectures delivered in India. Later at lunch, Prof. Fowler recalled, with a guffaw, she told him, "Prof. Fowler you can worry all you want about the nuclear reactions inside the stars. I have to worry about how to feed six hundred million people."

On another occasion he said, "I was travelling in a train when I got the news that I had won the Nobel Prize. I had presumed that Fred (Hoyle) had got it too. When I returned I discovered that he had been left out of the Nobel Prize. I immediately rang up Fred and told him that I would not accept the prize. Fred told me, don't be a fool. Go ahead and accept it."

Once I asked him, "Prof. Fowler have you ever thought about problems of society?" He immediately answered, "Yes". Then he paused for a few more minutes and said, somewhat regretfully, "No, I haven't. I have been far too involved in my work to think of anything else."

Prof. Fowler had received any number of awards, honors and honorary degrees, apart from the Nobel Prize, including the Medal for Merit by President Harry Truman in 1948, the Barnard Medal for Meritorius Service to Science in 1965, the G. Unger Vetlesen Prize in 1973, the National Medal of Science presented by President G. Ford in 1974, the Eddington Medal of the Royal

Astronomical Society in 1978, the William A Fowler Award for excellence and distinguished accomplishments in Physics of the American Physical Society in 1986, the Legion d'Honneur Award from President Mitterrand of France in 1989, the Life Time Achievement in Science Award of the B.M. Birla Science Centre in 1990. He was elected Member of the National Academy of Sciences in 1956, Member of the National Science Board, Member of the Space Science Board. He also received honorary degrees from the University of Chicago, the Ohio State University, the University of Liege, the Observatory of Paris, the University of Massachusetts.

It is a great honor to have been invited to deliver the Fourth B.M. Birla Memorial Lecture following in the footsteps of Fred Hoyle, Philip Morrison and Abdus Salam. I must express my gratitude to Dr. B.G. Sidharth, Director of the Birla Science Centre, for all he has done to make the arrangements for the travel here and the stay here of my wife and myself so pleasant and so comfortable. Finally we are most grateful to Mr. and Mrs. G.P. Birla for their gracious hospitality at their home and its beautiful gardens here in Hyderabad.

B.M. Birla was a very great man – an industrialist with great interest and participation in science, engineering and education. He was very public spirited and founded a number of institutions for the education of young and old alike. I have tried to think of an American of comparable stature and attainments to B.M. Birla and have decided upon Thomas Jefferson. Jefferson wrote our Declaration of Independence and was our third President. He was the owner of a large agricultural estate in the state of Virginia and managed it with close attention to details. In those days the workers on such estates were considered to be slaves, but Jefferson was kind and generous to his slaves in contrast to many other landowners at the time. Jefferson founded the University of Virginia and interested himself in science and invention. I am proud to tell you that we Americans had a B.M. Birla and his name was Thomas Jefferson.

Now I will turn to my subject for today. In this talk I will take you back eleven billion years ago to the first few thousand seconds after the origin of our universe of which we and the earth and the sun and our galaxy, the Milky Way, are but a very small part. Many cosmologists think my age of eleven billion years is too short and many prefer a number more like fifteen billion. We need not worry about this detail today.

The title of my talk should have been OUR EARLY UNIVERSE not THE EARLY UNIVERSE. Many cosmologists, and I am one of them, believe that our universe is just an expanding bubble in an otherwise infinite universe both in space and time. This infinite universe consists of strange stuff about which we know very little except that it has exceedingly high density. From the basic equations which Einstein gave us we also know that this stuff exerts negative pressure. It is equivalent to Einstein's cosmological constant. In the Friedmann/Elinstein equation for pressure in the universe the cosmological constant term is preceded by a minus sign. Thus instead of compressing our expanding bubble it actually maintains the expansion. Eleven billion years ago a phase transition took place which changed this strange stuff into ordinary matter like you and me which has been expanding ever since. There may be other expanding bubbles but we will never be able to observe them through the dense intervening stuff.

Now why am I taking you back eleven billion years to the first few thousands of seconds? I am doing so because it was during this short interval that the major part of the first four elements in the periodic table, hydrogen, helium, lithium and beryllium, was produced as well as a small fraction of the heavier elements. Most of the heavier elements were produced in stars but that is another story. From the early production of the light elements we can learn indirectly a great deal about our observable universe. How that can be is my story today.

Before continuing let me make a disclaimer. When one has worked as long as I have on my subject today, one comes to be considered an expert. Well, I am no expert so let me tell this story. I think it is fair to say that we look up to members of the medical profession as experts. Well, more or less. But you know how it is. When you are ill, you go to your doctor. He diagnoses your problem, prescribes treatment and you do what he tells you. He is the expert. Well some time ago I sprained my left wrist. It was painful so I went to my doctor. He took X-rays and found it was not broken and was just a severe sprain. Then he dismissed me. But as I was leaving his office he said, "I want you to bathe your wrist in hot water three times a day." I was flabbergasted. I said, "Doctor, my mother told me to bathe a sprain in cold water." "Well," he said, "your mother was wrong; my mother told me to use hot water."

Now I will return to my subject.

George Gamow, the great cosmologist, argued that the universe erupted in a gigantic primeval fireball from an initial state of very high temperature and density. Fred Hoyle termed it the "Big Bang," somewhat in derision, since he believed in a steady state model with no origin and no ending. Gamow's ideas were based on Edwin Hubble's discovery at Mt. Wilson that all the galaxies in our observable universe were receding from each other at enormous speeds. This was taken as strong evidence against a steady-state universe and in favor of a universe that was indeed expanding from a highly concentrated initial state.

Gamow's expanding universe was uniform, isotropic and homogeneous. It is commonly referred to as the standard big bang model. I call it the obsolete big bang model for the reasons I'll present later. In 1967 Wagoner, Fowler, and Hoyle calculated the abundances of hydrogen, helium, and lithium produced in the first thousand seconds or so at high density and high temperature. We, and later others, found agreement with observations on the abundances of hydrogen, helium, and lithium for the present mean density of ordinary matter like you and me in the universe equal to about 10% of the so-called critical density. The critical density can be calculated from Hubble's measurements.

It can be understood as follows. If the actual density is more than the critical density then gravitational attraction between elements of matter will eventually stop the expansion and reverse it to a contraction which will finally lead to a "Big Crunch". If the actual density is equal to the critical density the expansion will continue forever but with a velocity of expansion which will eventually equal zero. In order for this to be the case it is necessary for Einstein's curvature parameter for the universe to be equal to zero. The surface of the earth is curved in two dimensions of space. Einstein introduced the idea that the universe could be curved in four dimensions, three for space and one for time.

Einstein's curvature parameter is indeed equal to zero in a variation of the Big Bang model proposed in 1981 by Alan Guth of the Massachusetts Institute of Technology. In this new model it was proposed that a very small fraction of a second after the Big Bang, the size of the universe, prompted by the energy release associated with a breaking of the unification between the fundamental forces of nature, underwent a period of tremendous growth, increasing its size by a trillion, trillion, trillion, trillion times. A trillion is a million million. In a short time the expansion rate of the universe decreased dramatically and Hubble's relatively show expansion was recovered. This spurt in the growth of the universe is known as Inflation and is referred to as the Inflationary Model.

The Inflationary Model requires that the average density of the universe be equal to the critical density. Thus, if Wagoner and Hoyle and I were right twenty-two years ago, 90% of the universe must consist of some form of exotic matter. Elementary particle theorists have proposed many exotic particles in recent years such as axions, photinos, and WIMPS. Don't ask me what they are but I will tell you that W, I, M, and P are the first letters of Weakly Interacting Massive Particles. None of these exotic particles have been observed at high energy accelerators around the world up to the present time and the search goes on. I think it will be fruitless.

Gamow's Big Bang was homogeneous, everywhere the same in the universe. Fortunately the Inflationary Model permits the early universe after inflation and during Big Bang nucleosynthesis to be inhomogeneous with regions of high density immersed in a low density sea as first pointed out by Edward Witten of Princeton. Then James Applegate of Columbia and his collaborators and Robert Malaney and I at Caltech showed that Big Bang nucleosynthesis in an inhomogeneous universe could reproduce the observations in hydrogen, helium, lithium and also beryllium with the mean density of ordinary matter like you and me in the universe equal to the critical density ($\Omega_b = 1$). There is no need for exotic particles. That is the message of my lecture today. The theorists can ignore the vision of axions, photinos, and WIMPS as well as the sugar plums which dance in their heads.

These conclusions are illustrated in Table 1 which shows that, for $f_v = 0.11, \Omega_b = 1$ and $A_0 \geq 0.3$, as defined in the table, the abundances of H_2, He_3, He_4, Li_7 and of course H_1 are approximately given by nucleosynthesis in an inhomogeneous universe. Moreover Table 2 shows that the primordial

Table 1. NUCLEOSYNTHESIS IN AN INHOMOGENEOUS UNIVERSE WITH $f_v = 0.11$ AND $\Omega_b = 1$, Malaney and Fowler, Ap. J. 333, 14 (1988)

	Average Mass Fraction			
	H_2	He_3	He_4	Li_7
$A_0 = 1$	1.6(−5)	3.0(−5)	0.25	4.8(−10)
$A_0 = 10^{-1}$	6.7(−6)	2.2(−5)	0.25	1.5(−9)
$A_0 = 10^{-2}$	5.0(−6)	1.1(−5)	0.25	1.5(−8)
$A_0 = 10^{-3}$	4.7(−6)	6.4(−6)	0.25	2.3(−8)
No Diffusion	4.7(−6)	5.6(−6)	0.25	2.4(−8)
Observed Limits	>5(−6)	<3(−4)	0.22 − 0.26	2 − 8(−9)Pop I
				3 − 9(−10)Pop II
No Li_7 Problem for				2 − 8(−10)LMC
$A_0 \geq 0.3$				

Table 2. Be_9/H_1 IN OLD POP II STARS

STAR[1]	$log\, n(Be_9)/n(H_1)$
$HD134430$	< −11.9
$HD74000$	< −12.2
$HD19445$	< −12.3
$HD140283$ (Lowest observed value for	< −13.2
Be_9/H_1 produced in the Big Bang)	
SOLAR SYSTEM[2]	≈ −10.3
THEORY	
HOMOGENEOUS BIG BANG[3]	≈ −17.5
INHOMOGENEOUS BIG BANG[4]	≤ −13.0

abundance of Be_9 is also given by nucleosynthesis in an inhomogeneous universe [1–4]. The other parameters used in obtaining these conclusions are summarized in the final paragraph which follows.

A_0 measures back diffusion of neutrons into proton-rich region in which $Y_n^{(p)}$ would otherwise be small

$$Y_n^{(p)}(t) = A_0 Y_n^{(n)}(t)$$

$A_0 = 1$ for rapid diffusion relative to time scale for nucleosynthesis.
$A_0 = 0$ for no back diffusion.
$A_0 \geq 0.3$ yields mass fractions in agreement with observed limits.
Ω_b = baryon density/critical denslity
f_v = proton rich fraction of volume of the observable universe
$1 - f_v$ = neutron rich fraction of volume of the observable universe
$Y_n^{(p)}$ = mass fraction of neutrons in proton rich regions after back diffusion
$Y_n^{(n)}$ = mass fraction of neutrons in neutron rich region after back diffusion

Conclusion

And now my conclusion. What I have been telling you permits us to believe that we may well live in the simplest of all the universes compatible with Einstein's theories of special and general relativity. Its curvature parameter is zero, its cosmological constant is zero, its total energy is zero, its space-time is Euclidean, and its matter is stuff like us. I think Einstein would like that. I do, and I hope you do too.

References

1. S.G. Ryan, M.S. Bessell, R.S. Sutherland, and J.E. Norris, to be published (1990).
2. A.G.W. Cameron *Essays in Nuclear Astrophysics*, eds. C.A. Barnes, D.D. Clayton, and D.N. Schramm, Cambridge University Press (1982), p. 23.
3. R.A. Malaney and W.A. Fowler (1989): $\Omega_b = 1/40$, Ap. J. 345, L5 (1989).
4. R.A. Malaney and W.A. Fowler (1989): $\Omega_b = 1$, $f_v R \geq 10$, $R \equiv$ ratio of density in proton rich region to density in neutron rich region.

The Long-Term Future of Particle Accelerators

Simon van der Meer

CERN, Geneva, Switzerland

Fig. 1. Simon van der Meer with Mrs. G.P. Birla

B.G. Sidharth (ed.), *A Century of Ideas*, 27–41.

Simon van der Meer, was born in 1925, in The Hague, the Netherlands, as the third child of Pieter van der Meer and Jetske Groeneveld. He had three sisters.

Meer attended the Gymnasium in The Hague and passed his final examination in the sciences section in 1943. Because the Dutch universities had just been closed at that time under the German occupation, he spent the next two years attending the humanities section of the Gymnasium.

From 1945 onwards, Meer studied Technical Physics at the University of Technology, Delft, specializing in measurement and regulation technology. After obtaining his engineering degree in 1952, Meer worked in the Philips Research Laboratory, Eindhoven, mainly on high-voltage equipment and electronics for electron microscopes.

In 1956 he moved to Geneva to join the recently founded Centre Europeen de Research Nuclear (CERN). Here Meer's work was concerned mainly with technical design: poleface windings, multipole correction lenses for the 28 GeV synchrotron and their power supplies. In the meantime, stimulated by many contacts with people, understanding accelerators his interest in matters more directly concerned with the handling of particles was growing. Thus he proposed a high-current, pulsed focusing device (horn) aimed at increasing the intensity of a beam of neutrinos, then at the centre of interest at CERN and elsewhere.

From 1967 to 1976 Meer was responsible for the magnet power supplies, first of the Intersecting Storage Rings (ISR) and then of the 400 GeV synchrotron (SPS). Meer had invented an ingenious method for the dense packing of protons which circulate in an orbit in a vacuum chamber, guided by magnetic fields. One expected on theoretical grounds that the weak interaction is communicated by extremely heavy hypothetical particles, W and Z. In 1976 Carlo Rubbia presented an idea to convert an existent large accelerator into a storage ring for protons and antiprotons. The W and Z particles could then be produced in violent head-on collisions between the stored particles. Meer fine -tuned his method for use on the current of antiprotons. Rubbia's idea and Van der Meer's invention were combined in a large project and the first collisions in the CERN superaccelerator were observed in 1981. The discovery of the W and Z were announced in 1983 by Rubbia and collaborating large teams of scientists, basing the evidence on signals from detectors, specially designed for this task.

The Royal Swedish Academy of Sciences awarded the Nobel Prize in Physics for 1984 jointly to Professor Carlo Rubbia and Dr Simon Van der Meer for their decisive contributions to the large project at CERN, which led to the discovery of the field particles W and Z, communicators of weak interaction. However, he is ever ready to point out that he is an Engineer!

Van der Meer married Catharina M. Koopman in 1966 and they have two children Esther and Mathijs.

In 1990 Meer retired from CERN.

Prof. Meer has received numerous honors and awards including, Horzours Loeb Lecturer, Harvard University, 1981, Duddell Metal, Institute of

Physics, 1982, Honorary Degree, Geneva University, 1983, Honorary Degree, Amsterdam University, 1984, Foreign Honorary Member, American Academy of Arts and Sciences, 1984, Correspondent, Royal Netherlands Academy of Sciences, 1984.

---

1 Introduction

This talk is going to deal with the big machines that are used to do research on the smallest parts of matter and on the way they interact. In fact, at the bottom of all physical phenomena are particles and forces between particles. To observe these on an even smaller scale, we have to accelerate the particles to high energy. The reason is very fundamental: it is Heisenberg's uncertainty principle that is at the basis of all modern physics. The uncertainty in position, multiplied by the uncertainty is momentum, cannot be smaller than Planck's constant h. So, if we want precise position measurements to study small things, the momentum of the particles involved should not be too low, because this would imply a precise knowledge of momentum so that Heisenberg's principle would be violated.

Another way of saying the same thing is that each particle can also be interpreted as a wave; for looking at small details we need a short wavelength and this again corresponds to high momentum.

The only way that we know to accelerate particles to high energy is to use charged particles and to let them move in an electric field. If, for instance, an electron moves from a negative electrode to a positive one, it gains energy equal to the potential difference (volts) multiplied by the particle's charge. The energy is expressed in electron volts (eV); this is the amount gained by an electron moving across a potential difference of one volt.

Present-day accelerators attain energies of hundreds of billions eV (hundreds of GeV). Clearly, we cannot make such high voltages; much more than a few million volts is not practical. Therefore, we let the particle traverse successive accelerating gaps (Fig. 2). To avoid too high voltages, we make successive electrodes positive and negative in turn so that their voltages do not add up. However, as the particles move from one gap to the next, we change the polarity, so that the field seen by the particles always has the same direction. In this way we can get quite fast acceleration. For instance, 25 million eV per metre of accelerating structure can readily be obtained. The frequency at which the polarity is changed will be quite high; billions of times per second. The electrodes form a structure that resonates at this frequency.

It should be noted that the velocity of particles cannot be increased indefinitely. As soon as the speed approaches that of light, its increase becomes slower and slower; however, the particle's energy is still increased by the accelerator, and this is the quantity that matters.

A well-known trick to reduce the size of an accelerator is to let the particles move in circles by deflecting them in a magnetic field. At each revolution

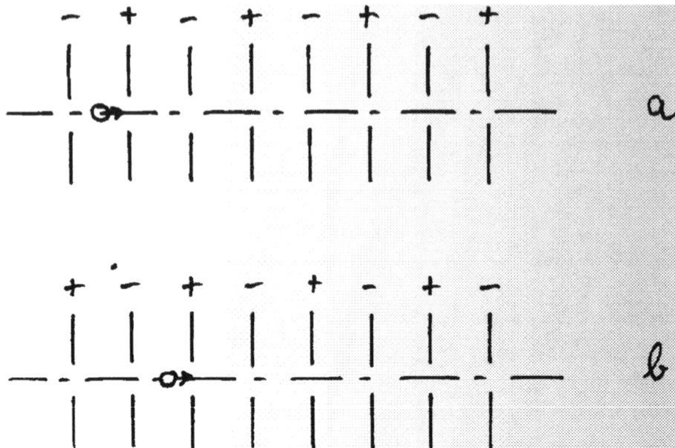

Fig. 2. An accelerating structure consisting of many gaps. The polarity is inverted as the particle passes from one gap to the next, so that it is always accelerated

the particles traverse one or more accelerating gaps so that their energy is increased all the time. If we want to keep the particles on the same circle, we also have to increase the magnetic field while they are accelerated. This is the principle of the synchrotron. Such machines are pulsed: the magnetic field and the particle energy increase during a certain time, typically a few seconds. The particles are then extracted and used for experiments. The magnetic field is reduced again and the whole cycle restarts. Most modern high-energy accelerators are of this type. Figure 3 shows an aerial photograph of CERN, the European particle physics laboratory, at Geneva. The accelerator rings are shown as giant circles; the largest one has a circumference of $27\,km$. It is built deep underground in a tunnel that passes below fields and villages, invisible from the surface. Figure 4 shows the inside of the tunnel, where one can see the long deflecting magnets. The curvature of the tunnel may just be seen.

The purpose of these machines is to make collisions between particles where a high energy is concentrated in a very small space. The energy is not high in absolute terms; in a particle collision made by our largest accelerator the energy liberated per collision is comparable to what a fly needs to lift one of its legs. However, it is concentrated in a very small volume indeed; this kind of concentration does not normally occur in nature. It is typical for conditions during the "big bang", a fraction of a second after the start of the universe. In fact present theories about the phenomena occurring at that time rely heavily on the results obtained at laboratories like CERN. I am, however, not going to speak about this, since I am not a theoretical physicist, but an accelerator builder.

Fig. 3. An aerial view of CERN, Geneva with its underground machines shown schematically

Fig. 4. Inside of the LEP tunnel, with deflecting magnets

2 Colliders

Until about twenty years ago, we studied particles and forces by shooting high-energy particles on stationary matter. They would then hit the nucleus of an atom, usually a simple atom like hydrogen, whose nucleus is just a proton.

The problem with this is that most of the energy given to the accelerated particle will turn up, after the collision, as energy of movement (kinetic energy)

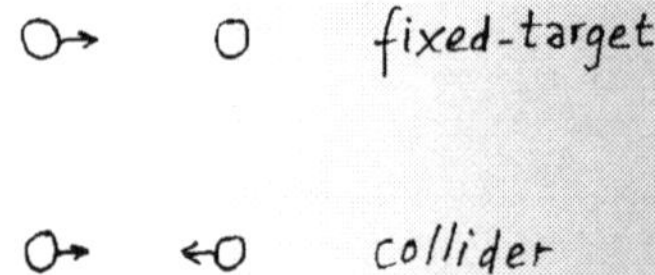

Fig. 5. Fixed-target vs collider. In the fixed-target arrangement only a small fraction of the particle's energy is freely available; the rest must be used for accelerating the interaction products

of the original particle and its target. This is because the total momentum of the two particles must be conserved. It can be shown that the amount of energy remaining for the interaction proper (and, for instance, available for creating new particles out of energy) only increases slowly with the incoming energy. As accelerators get larger this effect starts to limit their performance.

To avoid this problem, we nowadays usually observe collisions between two particles that have both been accelerated and that make head-on collisions (Fig. 5). Since they have opposite momentum before the collision, the total momentum is zero and momentum conservation does not require that any of the interaction products are accelerated much. So the entire incoming energy can be used for the interaction. We call such machines colliders (although, of course, the old "fixed-target" machines also produced collisions).

The disadvantage of colliders is that beams of accelerated particles have a density that is much lower than that of ordinary matter. So if two of these beams collide, the chance of a close encounter of two particles is quite small. We speak of "luminosity" as the property that determines the chance of particles to collide. In fact, the particles have a finite size (cross-section), and if we multiply this cross-section with the luminosity, we find the number of interactions per second.

It now turns out that the particle's cross-section decreases rapidly as its energy increases. The general tendency is for it to decrease inversely with the square of the energy. (This is a phenomenon again connected with Heisenberg's uncertainty principle.) Therefore, as the energy of colliders increases their luminosity must also increase rapidly. This presents a great technical challenge. So far, we have been able to meet it, but it becomes more difficult as the energy increases.

3 Electron vs Proton Machines

Most high-energy accelerators accelerate protons. Now protons, as we have recently found, are composite particles; they consist of three quarks, bound together by the "force particles" called gluons. As a result collisions involving protons tend to be complicated affairs; we are really interested in quark-quark collisions, but in most of the proton-proton collisions the quarks do not score a direct hit on each other. As a result, a lot of uninteresting "events" happen, producing a background to the few interesting quark-quark collisions. The cross-section for this background does not decrease with energy and is billions of times higher than the quark-quark cross-section at present-day energies. This makes experiments more difficult. Also, the individual quarks only have a fraction (about 10%) of the proton energy.

It would seem to be more profitable to accelerate electrons rather than protons. Electrons, like quarks, are elementary particles as far as we know. Their cross-section is comparable to that of quarks at high energy. The strong background from proton-proton collisions would not be present with electron-electron or electron-positron[1] collisions.

The reason that we still use proton accelerators is connected with a problem specific to electron machines. Charged particles will spontaneously radiate energy when deflected by a magnetic field. This energy loss counteracts the acceleration process. The effect depends strongly on the mass of the particle; while it is unimportant for protons, electrons suffer from it because they are about two thousand times lighter. As a result, we cannot deflect high-energy electrons too strongly. This is why our largest machine at CERN (with $27\,km$ circumference) is so large; it is an electron machine and therefore the deflection magnets have to be relatively weak. In fact, the energy is only $50\,GeV$, whereas the smaller ($7\,km$) SPS ring visible in Fig. 3 is for $400\,GeV$ protons. Nevertheless, the recently completed large ring (called LEP, i.e. large electron-positron machine) is, we hope, a fruitful physics tool because of the relatively clean experimental conditions.

Still larger rings for electrons would become very expensive. Moreover, the radiation problem increases strongly with energy. This is why LEP is probably the largest electron ring that will ever exist; for still higher energy we must use linear accelerators (linacs), where the radiation is not a problem (but many other things are).

The design of a linear collider of suitable energy (e.g. $500\,GeV$ or $1000\,GeV$) is very difficult. Several groups work on a design for such a machine: Stanford (USA), CERN (Geneva), Novosibirsk (USSR) and KEK (Japan). It is only fair to say that, even if the money would be available, we would at present not have a quite satisfactory design for such a machine. In fact, the next large accelerators will probably still be circular proton machines. However, in the long term we will have to switch over to linear electron-positron colliders. The rest of this talk will be devoted to an explanation of the difficulties that are encountered in designing such a machine.

---

[1] Positrons are the anti-particles of electrons; they have opposite charge and an electron-positron pair will annihilate and transform into energy, thus forming an ideal subject for collision experiments.

4 Linear Electron-Positron Colliders

In a circular machine, the particles turn round and meet the opposite beam again and again. In a linear collider, on the other hand, the particles meet the other beam only once and are then lost. This makes it so difficult to obtain a high luminosity. Also, for high energy, the machines tend to become very long.

One linear electron-positron collider exists; it is called SLAC and located at Stanford (USA). As Fig. 6 shows, it is somewhat special; a single linear accelerator is used ($3\,km$ long). Electrons and positrons have opposite charge, but can both be accelerated in the same machine because they pass at slightly different times, so that they see opposite electric fields (Fig. 2). At the end of the linac, the electrons and positrons are deflected through different, opposite, circular paths to the interaction point. This is still just possible at the energy of this device ($50\,GeV$). At higher energy the radiation loss in these curved

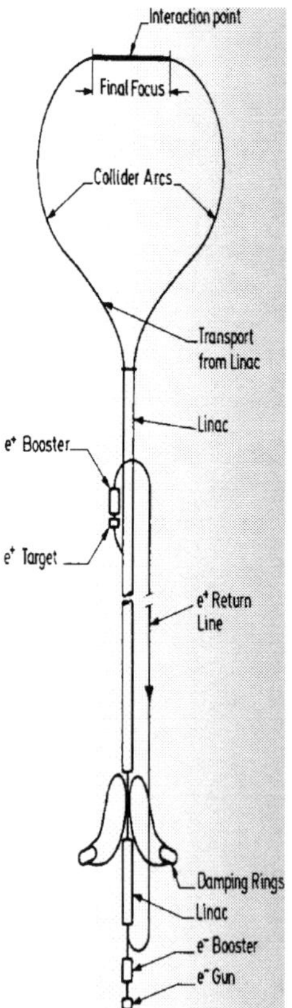

Fig. 6. Schematic arrangement of the SLC Collider at Stanford (USA). The linac is $3\,km$ long

parts of the machine would become far too high, even for a single passage. Therefore, future linear colliders must consist of two separate linacs, shooting the particles towards the interaction point from opposite directions.

The problem is how to obtain the required high luminosity. We try to increase this by concentrating the particles in short bunches (1 mm long, or less) with as many particles per bunch as possible. The number of bunches colliding per second cannot be too high because the power in each beam would become excessive. Typical bunch repetition frequencies would be somewhere between 50 per second (50 Hz) and a few KHz.

The luminosity L is given by

$$L = \frac{N^2 f}{4\pi\sigma^2}$$

where N is the number of particles per bunch, f the bunch repetition frequency and σ the transverse size of the beams in the interaction point. Increasing either N or f would increase the beam power, which tends to be a limiting factor in these machines. Clearly, we would gain much by reducing σ as much as possible. This means focussing down the beams to a very small spot in the interaction point.

5 Limitations to the Spot Sizes

Very strong magnetic lenses must be used to do this. A limitation, apart from the purely technical one of making strong enough focussing lenses, is also the fact that different particles will have slightly different energies. This causes slightly different focussing strength of the final lenses and as a result not all particles will converge to the same focal spot. This effect called chromatic aberration, can to a certain extent be counteracted by appropriate focussing schemes. Much work is going on in this field. Future machines may work with spot sizes of the order of 10 nm (1 nm = 1nanometre = a millionth of a mm). This is only about hundred times the size of a hydrogen atom. Clearly, the exact alignment of the two accelerators will be critical; after the kilometres of acceleration the beams must meet each other with a few nm precision. We believe that this problem although difficult, may be solved.

A further snag is that the intense beams may perturb each other at the focal point. In fact, it might be thought that even in a single beam the destructive forces might be quite strong because the charged particles will repel each other. However, fortunately the moving charged particles represent an electric current that creates a magnetic field around it. It now turns out that for particles with a velocity near to the light velocity the attractive effect of the magnetic field just cancels the repelling effect of the electric field.[2]

---

[2] An equivalent way of saying the same thing is that, for particles with near-light velocity, space and time are different from what is seen by a stationary observer

However, at the interaction point, because of the opposite movement of the two beams, the magnetic and electric field will reinforce each other. As a consequence, the electron and positron beams will attract each other strongly. This may be an advantage up to a point: the beam size will shrink because of the mutual attraction and this will increase the luminosity. However, above a certain limit, this "disruption" effect becomes so strong that the beams will be over focussed and spread out again before they have had much chance of interacting.

Moreover, in being so deflected by each other, the particles will again radiate energy. This effect (called "bremstrahlung") turns out to be so strong that the beams may loose too much energy during their interaction.

A third effect in the crossing point is the spontaneous creation of electron-positron pairs by the movement of the particles in the strong field of the opposing beam. These additional particles of lower energy may disturb the experimental equipment around the interaction point.

Finally, after the collision, the beams must be dumped. Since the power in each beam is high (typically a megawatt), the diverging beams should not hit any part of the opposite linac before being sufficiently diluted, to avoid excessive heating.

6 Beam Emittance

It turns out that all these effects limit the parameters of these machines in different ways so that there is finally only a limited choice. The conclusion one arrives at is that the beams must have a very small "emittance".

The emittance is the product of the transverse size of the beam and its angular spread. It may be shown that this product is not changed by focussing the beam: as its transverse size decreases, the angular spread increases. The emittance may, however, be decreased at low energy (before most of the acceleration) by letting the beams rotate in a small "damping ring". The radiation by the beams in this ring will lead to a reduction of the emittance. In Fig. 6 two small damping rings for electrons and positrons may be seen. The design of such rings has made much progress in recent years and, at least on paper, it seems that we could now build damping rings suited to the purpose, that would decrease the emittance typically to 1% of what is now obtained at Stanford.

However, it is not enough to reduce the emittance of the beams before acceleration. We must also conserve this low emittance over the whole length of the accelerator (i.e. 10 or $20\,km$). There are, unfortunately, several effects that tend to increase the emittance.

---

according to Einstein's special relativity. In the particle's frame, the accelerator seems much shorter and the time to traverse it is also shorter, so that there is no time for the beam to disintegrate!

First of all, the beams must be kept focussed all along the length of the accelerator, because they would otherwise become too sensitive to perturbing transverse deflecting fields. This is done by means of magnetic lenses placed at regular intervals. Unfortunately, even a very small misalignment of such a lens will cause a noticeable deflection of the beams that may easily become greater than the beam's angular spread. It might be thought that such a deflection could easily be suppressed at the end of the accelerator by compensating magnetic fields. However, the problem is that not all particles will have exactly the same energy. As a result they will be deflected differently and subsequently their orbits through the focussing lenses will also be quite different. As a consequence, a single deflection of the beam somewhere along the linac will cause the beam to spread out and finally increase its emittance (so-called "chromatic smearing").

Another disturbing effect is the so-called transverse wakefield. This is again caused by small misalignments of the beam with respect to the accelerating structure. The charged bunches will leave a high-frequency field behind; in fact, it is this wakefield that subtracts from the accelerating field and so enables the transfer of energy from the field to the particles. The wakefield is normally longitudinal, but due to misalignment it may also have a transverse component. The wakefield caused by the front of the bunch will then deflect the particles in the tail. The front, once it is misaligned, will oscillate transversely under the influence of the focussing fields. The wakefield will therefore also oscillate; and if the particles in the tail of the bunch have the same proper frequency of oscillation, they will be excited in resonance. This effect may lead to complete break-up of the bunches.

One way to suppress the effect is to have different focussing strengths for the front and the tail of the bunch. The CERN study group has shown that this may be obtained by using focussing lenses that are excited not continuously, but by a frequency equal to the accelerating frequency. As a result, the focussing field will change noticeably during the passage of a bunch so that the resonance between front and tail of the bunch is avoided. In fact, calculations show that with this arrangement the "chromatic smearing" discussed before may also be reduced; the transverse wakefields may counteract this effect.

7 Acceleration

It still remains to be seen how we can accelerate the beam rapidly enough to avoid an excessive length of the machine. To give an example, the Stanford machine uses an average accelerating gradient of $20\,MeV/m$. With this gradient a $1000\,GeV$ machine would become $50\,km$ long. Clearly, we need a higher accelerating gradient, of the order of $100\,MeV/m$.

This seems possible in principle, but it does tend to increase the power consumption of the machine. In fact, linear accelerators consist of sections of high-frequency accelerating structure (Fig. 7). Each section is a cylindrical

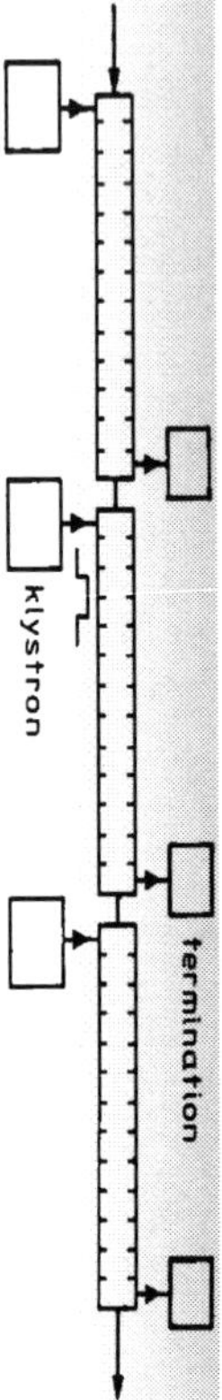

Fig. 7. Typical linac structure. Each section consists of gaps as in Fig. 1, powered by a high-frequency tube (Klystron). After the passage of the particles, the remaining electromagnetic energy is dissipated in the termination

tube, with so-called "irises": transverse diaphragms with a central hole through which the beam passes. The fields set up in this structure have the general shape shown in Fig. 2. The structure is "filled" with field by connecting one side of it to a high-frequency power source (transmitting tube, or "Klystron"). This source is switched on during a short time for each bunch. During this time (may be a few microseconds) the structure is filled with the high-frequency field. The bunch then passes and the particles are accelerated by the field. They should, however, not extract too much energy from the field, because the particles in the tail of the bunch would then see a lower field than those in front. Therefore, most of the electromagnetic energy contained in the field is still present after the particles have passed. This is then dissipated in the "terminating resistor" shown in Fig. 7. It cannot be conserved until the next bunch arrives; the power dissipation in the accelerating structure would be far too high for this. As a result, these machines consume far more power than is represented by the actual beam power. Increasing the gradient will aggravate this problem.

If we could make the entire accelerating structure superconducting, there would be no power losses and we could leave the high-frequency field on continuously. This would result in a large gain of efficiency. Unfortunately, present-day superconducting structures are quite expensive and do not support an accelerating gradient higher than about $10\,MeV/m$ – far too low for this application. It does not look as if this limit will be raised sufficiently in the near future.

One way of increasing the efficiency is to increase this frequency. The transverse dimensions of the accelerating structure scale with the wavelength, i.e. inversely with frequency. Therefore, for the same gradient, a high-frequency structure will contain less energy than a low-frequency one. We could, for instance, consider a frequency of $30\,GHz$ instead of $3\,GHz$ as used in the Stanford machine. The wavelength is then $1\,cm$ instead of $10\,cm$ and the accelerating structure will have a diameter of about $1\,cm$ only. Fabrication of such a miniaturized system may not be easy, but we think that we have little choice.

The problem here is that for this frequency no powerful sources exist. The Klystrons used for $3\,GHz$ cannot easily be scaled up to $30\,GHz$, since this would also imply a reduction of their dimensions and would therefore reduce their power handling capacity. a great deal of effort has therefore been spent on designing $30\,GHz$ power sources.

8 Generation of High-Frequency Power

The Stanford linear accelerator consists of about 900 sections, each $3\,m$ long, and each of them excited by a high-power Klystron. For a pair of accelerators with 20 times higher energy and working at $30\,GHz$ we would expect to need of the order of 100,000 sections of accelerating structure, each about $25\,cm$ long. Evidently, 100,000 Klystrons would represent an astronomical cost, even if we knew how to make $30\,GHz$ tubes of the necessary power rating.

Various schemes have been considered for replacing the Klystrons by something else. In fact, each Klystron is an electron tube, containing a low-energy, high-intensity electron beam. It would seem to be more practical (and it is at the basis of CERN's design for such a machine) to replace all Klystrons by a single low-energy, high-intensity electron beam running in parallel with the main linac. The high-frequency power can then be made by bunching this "drive beam" at the $30\,GHz$ frequency and letting it pass through structures similar to those of the main linac (Fig. 8). In fact, this "drive linac" will operate like the main one, but inversely: the beam energy is transformed into high-frequency electromagnetic energy instead of the other way round. This is simply achieved by letting the beam pass at the moment when the field decelerates it.

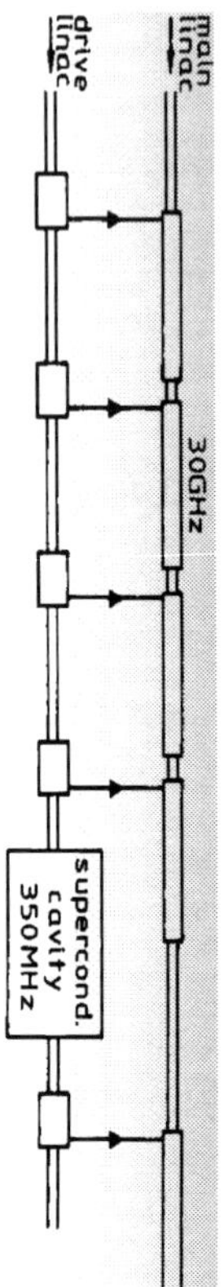

Fig. 8. Two-beam scheme for powering a linac. The drive beam is accelerated by superconducting cavities and decelerated by transfer structures that deliver power to the main linac's accelerating structures

A difference with the main linac is that the high-frequency structures of the drive linac will have a low "impedance" (large iris opening) so that the drive beam particles will lose much less energy than the main beam gains. Of course, energy is conserved; therefore the drive beam must be much more intense than the main beam.

The drive beam looses its energy to the transfer structures and must therefore be reaccelerated periodically. This can be done by highly efficient superconducting accelerating cavities; for this purpose their accelerating gradient is more than sufficient. These cavities would work at a much lower frequency than the main linac (350 MHz). Therefore, the drive beam would have to be bunched at this frequency, and each bunch should be subdivided into smaller "bunchlets" at 30 GHz for exciting the transfer structures. All this, although complicated, seems entirely feasible, the main problem remaining being to get the high drive beam intensity required.

Studies on a system for generating such a beam are in progress. Note that the drive beam, although relatively low in energy, would still have at least a few GeV; this means that the electrons in this beam have a velocity equal to the light velocity to within 10 parts per billion. They would therefore never be able to overtake each other and the bunches, once formed, would remain. Also, perfect synchronicity between the main and drive linacs would be guaranteed.

9 Remaining Problems

Before we could construct such a machine, it is clear that many of its features, so far only studied in theory, would have to be tested in detail. Fortunately, this machine, although long, would mainly consist of a large number of identical accelerating sections; these could be tested on a small scale before investing the large sums needed for the complete project. There are, however, several points still to be investigated.

(1) Manufacturing. We have to develop a low-cost way of producing the accelerating and transfer structures with the necessary precision (of the order of a few microns).
(2) Creating the dense bunches for the drive beam. Some designs exist, and work is in progress on a test facility, using a laser cathode and an initial accelerating gap with high gradient.
(3) Wake field effects, both in the main linac and in the drive linac. On paper, this problem seems near to solution, but the safety factors involved are not large and some more confirmation might seem to be desirable.
(4) Final focus design, including the disposal of the used beams without damage to the equipment of the opposing beam. Diagnostic methods to permit exact alignment of the beams with respect to each other must also be developed.
(5) Alignment. With the final microscopic beam size in the interaction point the alignment, especially of the final parts of each linac, must be highly stable and vibrations must be suppressed.

All these problems (and some additional ones) are being studied at CERN and elsewhere. We hope that this may lead to a realistic project somewhere at the end of this decade.

The cost of such a machine is difficult to estimate at the present state of design. However, it seems probable that great effort will be needed to keep it similar to the cost of the most recent CERN machines – of the order of one billion swiss francs. This might seem a large amount for a tool of fundamental scientific research without any practical applications. It should, however, be realized that per inhabitant of Europe this would correspond to the cost of a packet of cigarettes, hardly a great price for investigating some of the most fundamental aspects of nature.

Energy and Evolution

George Porter

Chairman, Centre for Photomolecular Sciences,
Imperial College of Science, Technology & Medicine, Department of Biology,
London, U.K.

Fig. 1. George Porter delivering the B.M. Birla Memorial Lecture

B.G. Sidharth (ed.), *A Century of Ideas*, 43–53.
© Springer Science+Business Media B.V. 2008

George Porter was born in Yorkshire in December 1920. After early education he went as Ackroyd Scholar to Leeds University. During his final year he developed an interest in Physical Chemistry and Chemical Kinetics. He also took a special course in Radio Physics, and later became an Officer in the Royal Naval Volunteer Reserve Special Branch which was concerned with the radar. The training which he received here in electronics and pulse techniques stood him in good stead in his later work on chemical problems.

In 1945 he went to Cambridge to work as a Post Graduate research student, where his first problem involved a study of flow techniques, of free radicals produced in gaseous photochemical reactions. This work led him to the use of short pulses of light, of shorter duration than the lifetime of the free radicals. He left Cambridge in 1954 and after a brief stint at the British Rayon Research Association he joined the University of Sheffield as Professor of Physical Chemistry in 1955.

Meanwhile he continued to work and showed how the flash-photolysis method could be extended and applied to many diverse problems of Physics, Chemistry and Biology. In 1966 he became Director and Fullerian Professor of Chemistry at the Royal Institution, succeeding Lawrence Bragg. Here his research group applied flash-photolysis to the problem of photosynthesis, extending his techniques to the nano second region and beyond.

Porter has several fellowships and honorary degrees to his credit. He was elected a Fellow of the Royal Society in 1960 and got its Davy Medal in 1971. He was the Liversidge Lecturer in 1969 and the President of the Chemical Society for many years from 1970, apart from several other distinctions. He obtained honorary D.Sc.'s from Sheffield, East Anglia, Utah, Leeds, Leicester, Heriot-Watt and other Universities. Apart from his numerous other distinctions, he got the Nobel Prize in Chemistry in 1967 which he shared with Manfred Eigen and Ronald G.W. Norrish one of his first teachers. He was knighted in 1972 and subsequently became Porter. He died in 2002.

Porter was interested in communication between scientists of different disciplines and between scientists and the non scientists, contributing to many films and television programmes.

He was a delightful personality with a zest for living. The day when he was to land in Mumbai on his way to Hyderabad, there were riots in that city and was advised to avoid visiting it. I spoke to him over the phone and he brushed all this aside saying that we shouldn't go by what the media and politicians say. He came to Hyderabad accompanied by his wife Lady Stella and thoroughly enjoyed the sights, a press conference and of course his lecture. He remarked that there was a lot more enthusiasm for science in the Indian media compared to the Western media. We had suggested a unique trip to the tea gardens of the North Eastern State of Assam, after his visit to which he readily agreed. Lady Stella had spent her childhood in India and the Porters visited a couple of nostalgic places which still evoked memories of the Raj. They were thrilled by the experience as also by their visit to the tea gardens.

Dr. Sidharth, ladies and gentlemen, distinguished guests; it is a very great pleasure and honor for me to be able to deliver this B.M. Birla Memorial Lecture, specially since I know that Mr. Birla was one of India's greatest philanthropists. He was an admirer and a lover of the ancient Indian culture. But he was aware and said that in the modern world it is necessary for India to also embrace the modern culture if it is going to flourish, the modern culture, which involves science and technology and in this way for this reason he very wisely and nobly, gave the support and money necessary not only for the Birla Science Prizes but he contributed greatly to the Science Centre. So for many other similar objectives we are all greatly in his debt for this. He was aware of the importance of the young people, the importance of all the people understanding science.

As Dr. Sidharth has pointed out, my predecessors in this lecture series have been a very distinguished four and I shall find it very difficult to follow them. But it is noticeable that they have all been in rather elated fields of astronomy and mathematical physics and so I am the first one who is not in the category, well I don't know what I really am. I started as a chemist and now am in a Biological department and so I am on the Chemical Biological side of science and so I hope this is some justification for my being here and balancing the science as I hope Mr. Birla would have wanted.

I have called my lecture Energy and Evolution, and that embraces Physics and Biology. I suppose that what I have in mind are the great things that have happened in the last 135 years since Charles Darwin; and the great problems that we have in this field today. In 1859 Charles Darwin wrote history on a grand scale and he gave mankind an intellectual shock which changed our concept of ourselves and our place in the world. Rather suddenly we have come to realize that the process of natural evolution which he described and which has served the world for three billion years may be about to cease or least to change in a profound way. The Darwinian changes of evolution occurred slowly, unnoticed by participants who had very little to say about the forms that their descendants would take. They merely flocked to survive and if they survived they had one privilege only and that was the privilege of handing on their genes. The situation has changed drastically in the last few years. One species, man now so dominates the earth that it is in his part to eliminate most of the other species if he so wishes. Those who do survive do so only because man finds them interesting and useful and he is busy with the natural evolution even of these. It is the end of the evolution, as Darwin knew it. Far greater powers to play God will soon be in our hands. Genetic Engineering will enable us to eliminate conquered genes and other unfavorable genetic information and even to change the nature of mankind. We may not wish to do this but it will become possible. What we see happening is a rapid transfer of responsibility for the future evolution into the hands of ourselves, the hands of one species, homosapiens. We are no longer pawns in the game of evolution. We are not even the kings and queens, we are the players.

Well, evolution like the rest of nature is governed by the science of Thermodynamics. Our great need is to have enough food and enough energy to survive. If we have those things we can probably provide the rest. But the science of Thermodynamics – some of you who are not scientists may find it a little difficult – is the one which frightens a lot of people. It is a lovely science really, I am not going to say a lot about it but one has to say a little about it. Because it is very close to our needs, we are all told to conserve energy. We are told that the conservation of energy is very important, the most important thing of all, by our political advisors who would one day no doubt pass a law about it and it will of course be called the Law of Conservation of Energy. But we have as you know, the first law of Thermodynamics. So what is the problem? It means that you can't get energy from nothing.

If all the energy being conserved means that you can't destroy energy and you can't create it – then that's alright. There would be no problem. But that isn't the only part of Thermodynamics. The second law tells us that you can't ever break even. You are going to loose some of your energy because energy is continually being degraded. And that is because the world automatically becomes disordered. There are just more arrangements possible in a disordered world than an ordered one and so naturally nature goes in that direction and that's the second law. We say entropy increases. So the world is running down over time, temperatures have become more even, we are eventually proceeding to a heat death. That is the second law.

That is alright until you begin to think about the local process, which has happened here on earth, the process is like Darwinian evolution which goes in absolutely the opposite direction. And that is a process of increase of order. It is contrary, isn't it, to the second law of Thermodynamics! To start with atoms and molecules and then with simplest cells, the prokaryotic and then the eukaryotic and then the multicellular organisms and then the fishes and the mammals and the particular evolution of man himself – it is a process of continuing order. Now how can that happen in accordance with the laws of Thermodynamics? Well it happens of course because we live on the earth, but we live with an important neighbor, the sun, which is a nuclear fusion reactor that works beautifully. It is a violent nuclear reactor. The sun's corona would engulf the earth many many times over. That nuclear fusion reactor fortunately, violent as it is, is situated ninety two million miles away, a safe distance to have a nuclear fusion reactor and it has provided of course all the energy, all the food, all the motivation for the development of everything on earth, particularly life itself.

It began by simple reactions in the atmosphere. The primitive atmosphere had no oxygen. It was an atmosphere of reducing compounds like Ammonia and Methane and water and Hydrogen. They were broken down by the ultraviolet light of the sun and turned into amino acids and similar compounds and eventually built up into quite complicated polymeric molecules which eventually were able to develop as living things. The process which occurs today which is responsible for all our energy, all our food, all our gas, coal, oil, wood, all our fuels is this wonderful process of photosynthesis, a process whereby the energy of the sun is absorbed by the green leaf and that solar energy is used to combine carbon dioxide and water from the atmosphere to make oxygen and food. The food would be the carbohydrates and sugar, starch and so forth and then the cycle is completed. We can wait, we can keep our food for a long time especially if it is a fuel because it could be fuel like oil or coal and then we can combine it again with the oxygen of the atmosphere and get the chemical energy out of the sun's energy. We can do this by burning the fuel at about 2000 degrees or we can do it by heating the fuel or food at about $17\,K$ which is the temperature much better suited for digestion and which is made possible by the use of enzymes as catalysts. So that's the magic cycle of life. Because once you have the energy the rest can happen in many many ways.

Now I want to look at the history of that energy with the history of photosynthesis – past, present, and future to some extent. The earth is about 4.6 billion years old. Life or the processes of developing life, have taken a great proportion of that time. Chemical evolution and these processes began almost immediately, presumably, and certainly there were biological processes occurring within a billion years of the birth of the earth. We know these things from micro-paleontology where one gets not fossils of dinosaurs and things, but rather fossils of the tiniest cells which go back to three billion years and probably before that. But certainly no later than two and a half billion years ago the process of photosynthesis began, probably using chlorophyll as it does today. So that's the long history. In recent history things have changed enormously. Look what's happened in the last one hundred and fifty years. A hundred and fifty years ago, the energy the world used was 90% wood. It was much more than that in countries like India but it was 90% wood in most of the developed countries. And in fifty years that had changed so that 70–80% was coal.

Finally a hundred years ago or so, coal began to be replaced by oil and gas which now dominate – from around 1970 the process has gone a little further and coal is rapidly becoming less and less popular. Let us look at that on a somewhat longer time scale, from the beginning, from the birth of Christ and hopefully I have taken it to 3000 AD. Hope somebody will be around that time to check whether I am right or wrong. You see we live in a very interesting time. We live absolutely on the top of the oil lake. This is not going to last very long from now. The oil would be gone in fifty or sixty years, well most of it. So what's left is very expensive. Coal would last quite a bit longer.

This period is just a blip in history, it is nothing in the long-term history. Coal, oil and gas have done a lot for us. We have been lucky enough to live at this time. We have enjoyed enormous prosperity on an average and that average of course has a big spread of individual participation in the joys of the fuel and energy.

So the problem is not only that the fuel is used up, it's a limited amount. It would soon be gone. Because of the other developments which have taken place as a result of energy and as a result of science and technology, the average

age of human life now is more than double what it was at the beginning of
this century. Now it is seventy six, beginning of the century it was thirty six
and that has meant that the death rate has gone down to the point where
there is an explosion of population and that population has to be fed and
fuelled. There is no problem about food because of the green revolution and
the development of insecticides and new species. The problem about food is
distributing it and getting it to the poorer people. There is about ten percent
more food produced than can be consumed at the moment. As you know in
Europe and many countries the politician's main job is how to stop people
from growing food. The whole thing has absolutely changed. Farmers are paid,
subsidized, not to grow food. So there is no food problem except as I said the
distribution one.

The fuel problem is quite different. The fuel problem is huge and insoluble,
almost insoluble. Our population in the last hundred years has gone from 1.49
to 5.32 billion people. The traditional use of energy, that is from wood and
dung and so on, has gone down slightly, but the industrial use that is from
coal and the making of electricity and so on has multiplied enormously and so
also the total world energy demand. Actually energy production has gone up
fourteen times and it will continue to do so because the population is going
to increase to ten billion by the year 2030. So we are going to have terrible
problems from about the time that oil and gas would be in short supply. So
there's the problem. I just mention here that one tends to give these figures
for the western world or for the average and I want to point out the enormous
difference between various countries. Whereas the average consumption in the
world, of wood and dung is 6%, for western Europe it is only 0.7%. In Africa
and India and many Latin American countries and so on even this is going
up to enormous proportions and as you know in rural India it is almost the
main fuel. But that is running out too. The trees are disappearing just as the
oil is disappearing.

I submit that there are really only two solutions to these problems in the
long term. We won't argue about when the fossil fuels would be gone, coal
would be gone, we won't argue about that. They will go eventually – that
is obvious and it won't be a very long time from now. What sources are we
going to have available then? In fact, one – nuclear fusion. A nuclear fusion
reactor hasn't yet been developed, but the other type of nuclear fusion is that
reactor I have already referred to as being at a safe distance of ninety two
million miles, and that is the sun.

That is the way that evolution and energy have developed for three billion
years, it is that process which has supported Darwinian evolution and one
asks why we can't go on in the same way as we always have by using the sun
as our source and that's what I want to devote the rest of my lecture to. First
I will talk a little about some of the research on how it works and so forth just
because it is interesting and then at the end we will have a word about what
this would mean if we can improve the efficiency of photosynthesis, what this
would mean in terms of the amount of land surface required and so on, to

supply the whole world with its energy. Nuclear power from nuclear reactors is a good possibility. I mean if we can do it. But it has problems as you know, disposable wastes, safety and so forth. We don't know which way do we go and I think to have all our eggs in one basket, to have any one source of energy in mind for fifty to hundred years time when the fossils fuels are gone would be well, it would be rash.

Let's look at the possibility of continuing to use solar energy as we always have via photosynthesis. Well let's have a look at the machinery. Take a bit of leaf, consider one cell. The chloroplasts absorb the light and are the engines, the green engines of photosynthesis. If we look at one of these under a microscope, electron microscope, we see one chloroplast and in this chloroplast there are membranes, they are lipid membranes, fatty membranes, they are two molecules thick with the oily ends sticking into each other. The molecules are all in chains, with the heads which are hydrophilic as they say, they like to be wet, they like to interact with water. That forms a scaffolding of the photosynthetic unit and it supports the lump like, elaborate and beautiful structures.

In the green leaf there are the proteins, these are again lipid molecules and this illustrates the hope for us which is occurring in the green leaf. There are two units called photo-system 1 and photo-system 2 which act in series to push the electrons across the membranes, the light is absorbed and electrons are pumped across the membrane. They reduce, they go on to a molecule which is called a quinone and go through cytochrome and through plastocyanin and they are pumped up into a higher energy and on to the compound in ADP. Now at one end then we have electrons which are reducing and which reduce carbon dioxide to carbohydrates eventually. Whereas at the other end we pump hydrogen off water and we get oxygen. So that is splitting water into oxygen and hydrogen. I will be looking at this process, which produces the oxygen in a little more detail. Now there are centers doing two things. First of all they collect, they have one centre at about three hundred molecules of the green chlorophyll across the cavity. These collect the light, they absorb the light and collect it and the energy moves around amongst these two hundred to three hundred molecules, just walks around until it comes to a trap. This is the special chlorophyll molecule which absorbs at $680\,nm$. So it is called a $P680$. Then the chemistry starts – the molecule is highly energetic, because it has taken the energy from an antenna, which collects the energy, that exotic trap of the electrons across the membranes as I have said from the donor to the acceptor, the donor being the water, the acceptor being the carbon dioxide. That's the process that occurs.

Now five years ago for the first time the reaction centre of photosynthesis – it was the reaction centre of a photosynthetic bacterium which was rather simpler but nevertheless it was a whole reaction centre – was crystallized. Well this happened about eight to nine years ago but the structures, the X-ray structure they crystallized was determined about five years ago by Hartmut Michel and Diessenhofer and it was one of the quickest Nobel Prizes they have ever got. They got the Nobel Prize the following year for having done that beautiful piece of work. They determined the biggest molecule whose structure has ever been determined or had been determined at this sort of resolution: The position of every atom is resolved at 2 angstroms.

They are but the scaffolding to hold chlorophyll molecules including those equivalent to the $P680$, the double molecule of chlorophyll, or two, three, or five chlorophylls with the magnesium missing and quinone, iron atoms and so on. So that is the beautiful structure which was determined and what was so exciting about this was that already although the structure had been looked at by X-ray crystallography, already it had been worked out what this structure was going to have to do and not only that but also how fast it did it and using fast reaction techniques it was shown – remember what this is doing, it is pushing the electrons across the membrane – it was shown that the electron goes from the double chlorophyll molecule to the chlorophyll molecule in four Pico seconds and from that many quinones in 200 Pico second and then to the iron in 10 milliseconds.

Now how do we know that? Well, this is an area I am not going to be able to go into detail. It is my own subject of research which is to use very brief flashes of light which enable us by then following them with a second probe flash to find out what happens. You get a movie sequence and the steps in that movie sequence may be only one Pico second apart. Now for those who are not familiar with these very short times, a milli second is a thousandth of a second, a micro second is a millionth, a nano second is a billionth, a Pico second is a millionth of a millionth and then we should eventually get to a femto second which is a thousandth of a million of a millionth of a second. This has been made possible by the laser, which was discovered in 1960, and now we are down to the very shortest times, which are possible. There is an Uncertainty Principle, which tells us that we can never get in Biology, information in less than about a femto second. However these are slow, you see these are Pico seconds only, a millionth of a millionth of a second and it is possible to trace them, but the exciting thing to anybody in this field is when having worked out this, Michel and others did this structure. Those molecules are at the right distance and in the right position. The green leaf is at least twice as complicated. The chemistry which occurs as I have already mentioned, in the green leaf, is the splitting of water into oxygen and carbohydrates by combining the hydrogen with carbon dioxide. You can combine these two and make hydrogen instead, but normally one makes carbohydrates. The leaf is as I said twice as complicated as the bacterium because it has two photo systems which I mentioned which act in series and we pump the electrons through photo system 2 and then to the photo system 1. Photo system 2 is very interesting because it splits water and makes oxygen and that's a very difficult thing to do and I want to say a word about the amazing work, some of it unpublished which we are carrying out on the photo system 2 and by the technique of flash photo systems which Dr. Sidharth referred to, which I have been doing for a long time. This has only just reached the very fast times of some Pico seconds. Photo system 2 consists of something like this.

There are three hundred chlorophyll molecules, which are the antenna which gather the sun's light and these are in a "light harvest" complex tool that contains most of the chlorophyll. We understand it fairly well. We want to study what happens at the heart of the reaction centre. So we have to ask our chemical friends to get rid of everything else and to leave us with just the centre where the real action takes place, getting rid of the light harmones, the light harvesting stuff. The Manganese is very important to make the oxygen but we know nothing about it. It makes things too complicated. We are going to get rid of the quinones as well so we can see what happens and so finally we are going to finish with just the reaction centre itself where the electron transfer and energy transfer goes from its double chlorophyll molecule to the single chlorophyll molecule.

This is the sort of measurement and result which is now possible. The time scale here is of the order of seventeen femto seconds. What we do here is we have a series of pulses which repeat over and over again to get better accuracy. You split that series of pulses into two and you use one half to excite the chlorophyll or the centre and then you go away for a few femto seconds and come back and use the second half to measure what has happened. You can note the difference between the two halves by making the second half go for a little trip up and down the room. It travels a third of a millimeter in a Pico second. So you just have a mirror which can be lengthened so that the time can be whatever you like. You in fact get a movie with one shot every seventeen femto seconds. So this is a precision which is possible now with those reaction centres. It is possible to show that the energy transfer between the chlorophyll's occurs in 100 femto seconds which is about the fastest event in Biology, which has been recorded. It is pretty well the fastest thing in Chemistry as well.

So it is fascinating but of what use is it? Well I think most scientists believe that if one can understand a process, a natural process, then you can do something about using it. You can find some use for it. You can modify it and so the first aim is to understand those processes which occur in the green leaf in the way that I have just sketched out for you. But now suppose we understand perfectly what happens in the green leaves, most of those protein complexes have been sequenced that is to say we are to know the order of the amino acids in the proteins, we know the order of the nucleic acids in the DNA and so forth. So genetic engineering becomes a possibility if we understand what it is we want to engineer. Suppose we can improve photosynthesis efficiency, what then are the possibilities for doing what I started to talk about, that is producing renewable energy sources from photosynthesis for the whole of mankind? I have to go back to Thermodynamics.

We started with Thermodynamics, we will finish with it. We must go back to Thermodynamics and ask what is the problem of maximum Thermodynamic efficiency that we can get in photosynthesis or in fact is it just the same in any other process like photo voltaic cells. What is the maximum efficiency we can get? The sun is a white light source. It is a body whose temperature is $6000\,K$ on the surface and so if you use well-known Thermodynamics, the Carnot cycle, you can very quickly work out that the efficiency of light bases alone would be 95%. The Carnot cycle says the efficiency is the difference in temperature over the higher temperature. The temperature of the sun is $6000\,k$, the temperature of the earth is $300\,K$. So that is $5700\,K$ over $6000\,K$ which is about 95%. So far so good but I am afraid that this is not the whole story. The efficiency is much less.

In the first place you are only going to absorb certain parts of the sun's spectrum. If you have something which absorbs one wavelength which is what usually happens, all of the other wavelengths are going to be wasted. The sun's spectrum is a bit scattered by the time it arrived on the earth. We can see the absorption bands of carbon dioxide and water which we hear so much about, which produce the green house effect. So very little is absorbed. And that's the main loss. We can't use all the sun's light efficiently. This is 95% that I have just said which is due to the entropy loss, the entropy creation of the radiation. But, the sun's real Thermodynamics temperature is not 6000, but much less. The reason is because the sun is a small disc, it doesn't go right round the whole celestial sphere of 4π. It is a small disk subtending a fraction of this. This reduces the efficiency and the temperature.

Another 18% loss in efficiency is due to the scattered radiation, the entropy created by scattering. There is a little difference there and I won't refer to it. But the big loss is due to the fact that the sun has infra red and ultra violet radiation which are not used and so the total efficiency to come to the crunch of the matter, the maximum possible efficiency is twenty seven percent. That's good and in fact 27% has been reached with photovoltaic cells. Exactly all these things apply to photovoltaic cells just as much as to the leaf. They are all exactly the same Thermodynamics. So 27% is what one can do. The best that can be done in the field is about 1%, the best that can be done in laboratories is about 3%, what is actually done across agriculture is about 0.3% and so we have a long way to go.

Suppose we could improve this 2% in the laboratories to 10% which is only about a third of the way to what is theoretically possible, 10% efficiency of photosynthesis? Then we have to ask the question: With what is feasible now is it feasible to produce all the fuel that mankind wants, all the energy that mankind wants simply by photosynthesis, by agriculture? This would be a wonderful thing for many countries in the Third World, many of which don't have fossil fuels nor the capital to produce nuclear energy. This is a thing that can be done on a big scale. If you can do it on an acre, you can do it on a thousand acres and so it is a thing that just one farmer can do on a small scale. So it's very interesting from the point of view of many countries like India, I suppose. Alright, suppose we could have 10% efficiency – we can do it for 5% if you think that ten percent is being too optimistic. We now have to ask how much sunlight there is, how much energy from the sun there is. Take

India – I am giving you the figure of the average amount of energy falling on 1 square meter, the average over day and night, winter and summer. The peak is 1000 watts per square meter but for the most part it is 300 to 100 watts or so. Two hundred and twenty five is about the average for the whole of India. Two hundred and twenty five watts per square meter.

So now then how much land would we need if we have that amount of energy per square meter and we have only 5% efficiency which is only two and a half times what we can do in the laboratory now? There is a bit of genetic engineering of the plants to be done and so forth. Surely it should be possible to up the efficiency to develop something of this level. The amount of land we would need to supply all means of energy which is a very important requirement is only 4% of the amount of land that can be used for mankind's food. It could be used in the deserts or forests, if one could develop the appropriate fertilizer to supply the plants in that condition.

So that I hope is some justification for all the fun I get out of my research. The most important process of life, the process by which our energy is collected by the plants and transmitted to the animals as food, is wonderful Physics. It is wonderful Chemistry as well – the energy and electron transport which occurs in the leaf. But it is nice to think that when the time comes as it surely will, there is a possibility of developing a clean and renewable source of energy. Let me just emphasize that this is totally clean. As I have said, the carbon dioxide this puts out is taken out of the atmosphere and it is a renewable cyclic process. But on the contrary when you burn coal or fuel you don't put the coal or fuel back – you put the carbon dioxide into the atmosphere entirely. Let us not worry at the moment about the green house effect. If there is a solution to the green house effect this must surely be it.

Thank you very much.

The Wonders of Pulsars

Antony Hewish

Cavendish Laboratory, Cambridge, U.K.

Fig. 1. Antony Hewish delivering the B.M. Birla Memorial Lecture

Antony Hewish was born in Cornwall in May 1924 to a Banker. He was the youngest of three sons and grew up in Newquay on the Atlantic Coast. Here he developed love for the sea and boats. After education at King's College, Taunton he went to the University of Cambridge in 1942. He was in war service at the Royal Aircraft Establishment in Farnborough during the period 1943 to 1946. He was also involved with the Telecommunications Research Establishment at Malvern. It was at this time that he met Martin Ryle.

Back at Cambridge, he graduated in 1948 and joined Ryle's research team at the Cavendish Laboratory, obtaining his PhD in 1952. He became a Research Fellow at Gonville and Caius College. In 1961 he transferred to Churchill College as Director of Studies in Physics. He remained here in various positions till his retirement as Professor of Radio Astronomy in 1989. Meanwhile in 1977 he took over the leadership of the Cambridge Radio Astronomy Group and subsequently headed the Mullard Radio Astronomy Observatory from 1982 to 1988.

B.G. Sidharth (ed.), *A Century of Ideas*, 55–64.

He started doing research on propagation of radiation through inhomogeneous transparent media. The first two radio stars had just been discovered and Prof. Hewish realized that their "twinkling" could be used to probe conditions in the ionosphere. Subsequently he developed methods to make the first ground based measurements of the solar wind, which were later adopted in the USA, Japan and India for long term observations.

Working with a new antenna, starting 1967, observations finally led to the discovery of Pulsars – which Prof. Hewish modestly describes as a stroke of good luck.

Prof. Hewish has received a large number of honors and awards, these including the Hamilton Prize of Cambridge in 1952, the Eddington Medal of the Royal Astronomical Society in 1969, the Michelson Medal of the Franklin Institute in 1973, the Hopkins Prize of the Cambridge Philosophical Society also in 1973, and of course the Nobel Prize in Physics in 1974 for his discovery of the Pulsars. He has several honorary ScD.s and is a Fellow of the Royal Society, the American Academy of Arts and Sciences, the Indian National Science Academy.

Prof. Hewish is a lively and delightful person. The years have not diminished his enthusiasm to lecture, teach and spread knowledge. One of the striking features of his personality is his humility and open mindedness. I have benefited from enjoyable metaphysical discussions as well.

It is a great privilege for me, it is a great pleasure and an honor to be invited to give the B.M. Birla Memorial Lecture. Clearly he was a very great man. We know that he was a captain of industry, a frontline banker. He was a man of vision. That shows because one can see his interest in recognizing that science is a part of human culture, that efforts should be made to present science in a manner that is understandable by the general public. It is I believe an obligation of scientists to undertake in this activity because it is true that science is a part of culture. But unless it is correctly presented it can be difficult to assimilate and this is a great loss to the general public. So I am very glad to be here to help in this activity of the B.M. Birla Science Centre to engage in trying to project some of the excitement and thrill of scientific research.

The discovery of Pulsars in 1967 was for me a great surprise because the research I was then doing was not aimed at discovering a new kind of star. I was then engaged in the branch of Radio Astronomy, which was to do with the study of quasars. In the early 1950s it was discovered that there are galaxies that emit powerful radio waves. Often they are only detectable by the radio waves, not by their optical emission and these radio galaxies were giving us new information about the past history of the universe. Quasars are a particularly active type of radio galaxy in which enormous quantities of energy are produced in a small volume at the centre and understanding them was a key problem. But we did not always know which radio galaxies were quasars. By then there were many hundreds known but radio telescopes were rather poor in those days at imaging radio galaxies and we had little idea how large they were, or whether they contained these active centres which are typical of quasars.

I had an idea that we could use a very well known phenomenon. We are all aware that as we look at stars in the sky they twinkle because of fluctuations in the earth's atmosphere. A discovery that we made at Cambridge in 1964 was that some radio galaxies also twinkle. They twinkle because the sun is blowing gas into space. We call it the solar wind. The solar wind fills our solar system and the irregular volumes of gas leaving the sun pass across the line of sight from the radio telescope to a distant galaxy and cause fluctuations in those radio signals. So some radio galaxies twinkle rather like stars twinkle but only if they are very small in angular size. If they appear as it were as ping pong balls in the sky, they would twinkle. Larger radio galaxies would not twinkle. I was planning to use this phenomenon to know which radio galaxies were quasars.

To do this well I had to design an entirely new kind of radio telescope. It had to be extremely sensitive to radio signals. It also had to be operating on a rather long wavelength which would generally be regarded as being bad for Radio Astronomy. The radio telescopes which were then fashionable used shorter wavelengths to achieve high angular resolution and the sharpest images. We built a new kind of instrument working at a long wavelength that was particularly sensitive to fluctuations.

The antenna is spread across an area of ground which exceeds four acres. That makes it a large instrument. A photograph looks more like agriculture than science. It is a field full of electrical receiving elements and there are 2048 of them making it an extremely sensitive radio telescope. It has another property, not shared by other radio telescopes, in that it can observe the whole sky in $24\,h$. It looks in many directions at the same time because it was necessary to observe the sky frequently in order to study the phenomenon of this twinkling which I mentioned. The survey began in 1967 and went extremely well and we found many twinkling radio galaxies. But there was one indication of something strange on the records.

My graduate student Jocelyn Bell who was a very diligent student, was responsible for the detailed analysis of the recordings and she showed these to me. In one direction there was a fluctuating signal which was typical for a twinkling radio galaxy, but the strange thing was that it was showing this effect at a time we did not expect. The radio telescope was pointed in a direction opposite from the sun where you do not usually see very much of the twinkling effect. Another puzzle was that the signal was not always present. The observations required repeated observations of the sky. We needed to observe the sky once a week and on some weeks we saw it and other weeks we did not. Now galaxies don't turn on and off like that! They are enormous objects, so something strange was going on. We decided to take a closer look by using a recorder that ran much faster so that we could get much higher time resolution and we saw a radio signal which was coming in the form of flashes of just over one second intervals. It looked very artificial like some navigating beacon in space.

This was a kind of radiation entirely unknown to astronomers. It took me a while to believe that these signals were actually real but after repeated, daily observations and determination of the radio bandwidth it was clear that the signals were coming from a unique direction in space located far beyond our solar system. They were not man-made radio interference. The only thing I could think of as a satisfactory explanation was that we were picking up signals caused by some alien intelligence elsewhere in the galaxy, because no astronomical objects then known could possibly produce flashing signals of this kind.

Now where do you expect life to emerge? Well it needs energy and it needs an environment in which chemistry similar to our own on this earth must be present. So we think that intelligent life will develop on another planet orbiting some distant star. The pulses kept time to better than a millionth of a second so I could check whether the signals came from a source in planetary motion. If an object emitting pulses is moving in an orbit, when it moves towards us the pulses gets compressed while when it is moving away the pulses get spread out in time. We call this the Doppler Effect in Physics and we can use it to determine the motion of the emitter. It took me three weeks to do the observations and I found absolutely no motion whatever. So I had to think of another explanation. From the short duration of the pulses I knew that the source could not be larger than a small planet. It was a relief to me that this object probably was not an alien planet because that would be a worrying discovery. But I don't want to elaborate how we were thinking in those days. I like to go on into science.

It was a wonderful discovery but you must find some theory before publishing it. So I began to think about very tiny stars. It was predicted in the 1930s that you could have tiny stars, only a few miles across, made of neutron matter which is enormously dense. A tiny star can vibrate very fast and vibrating stars are well known in astronomy. I thought that one of these tiny neutron stars might vibrate fast enough to generate flashes as it pulsated in and out. When we published the results it generated a sensation amongst astronomers all round the world. All sorts of theories were put forward, but after about one year it became generally agreed that the best theory was a rotating neutron star.

If you have a spinning star with beamed radiation, like a light house, then it produces regular flashes and we now believe that neutron stars behave like this. Only neutron stars can spin fast enough without breaking up. To understand the nature of these most unusual stars I need to discuss the basic physics of matter, under extreme compression.

In a star like the sun we understand what's going on inside. The sun is a huge ball of material which is mostly hydrogen and near the centre hydrogen is converted to helium in a nuclear fusion reaction where the temperature is high. Now ultimately, in millions of years, the sun must run out of nuclear fuel.

It must burn out and the question is what will happen then. Eventually the sun will cool down and it will be compressed by its own gravitational forces. The sun is presently inflated by gas pressure from the nuclear reaction at the centre. If you turn that off gravity simply must pull inwards the material of the star. So the question we have to ask is how ordinary matter behaves if we exert extreme pressure on it?

To understand this we must remember that ordinary matter is composed of atoms in which electrons are in orbit around individual nuclei. In solid materials as we have them on our earth, atoms are pressed together until the electron orbits begin to overlap. But if we squeeze matter hard enough we begin to crush that structure of ordinary atoms. We would then get what is called degenerate matter. According to quantum theory, if we put energetic particles into a small volume and compress them we increase their energetic motion. They rush around very fast because the waves which represent the particles have to fit into a smaller volume and the shorter the wavelength the higher the energy. Eventually the electrons which were orbiting become freely moving. They just dash around between the atomic nuclei and move randomly. We find this material inside White Dwarf stars which are highly compressed stars, about the size of our earth, which are the remnants of burnt-out stars. Then the matter has a density of about 1 tonne per cubic centimeter. But there is a further process that can take place. We can squeeze further until we crush matter into a new form altogether. That happens when the electrons and protons – the positive and negative charged particles of normal matter – combine to form new particles called neutrons which have zero electrical charge.

This is just the reverse of what happens under normal pressure when neutrons explode into electrons and protons with an enormous release of energy. In neutron matter a hundred million tonnes of material are contained in – about the volume of a teaspoonful – a cubic centimeter! Ultimately, when a star has burnt out, gravity can crush it into the form of a neutron star which would then be only a few miles across but nearly as heavy as it was in the beginning.

Now the evidence that pulsars were actually neutron stars came from careful observations and one of the best confirmations was the discovery of a pulsar inside a well-known optical nebula – the Crab Nebula. In 1054 oriental astronomers saw a star suddenly appear in the sky – and it was visible as a bright star for many weeks. Today we see in that direction a nebula. If we suddenly collapse a star like the sun that crushing is a very rapid process. The material of the star moves inwards at speeds approaching the speed of light and there is an enormous collision at the centre which ejects vast quantities of energy in shock waves. These blow the outer layers of the collapsing star into space which we see as a supernova explosion. The Crab Nebula is still expanding at the rate of several thousand miles per second.

Near the middle of the Crab nebula is a visible star which astronomers have long believed to be the remnant of the explosion. If it is a spinning neutron star it should be a Pulsar. No one looked for this effect until after the Pulsar discovery. Using radio telescopes a flashing Pulsar was discovered and with optical telescopes it was later found that the visible star was also flashing on and off in synchronism.

In the Crab Nebula the flashes come roughly thirty times a second. It is very rapidly spinning, which is expected for a young Pulsar. The rotation should be slowing down because the star is loosing energy. It was observed this flashing was, indeed, slowing down at exactly the rate predicted for a neutron star created at the time the supernova explosion occurred. So we have very powerful reasons for believing the neutron star theory.

Now how do Pulsars radiate? I wish I could give you the answer. This is still an open question and all I can do is to show you the ideas which currently are suggested. Normally a star produces radiation because it is hot. Pulsars are indeed hot objects. But this would not generate enough radiation to be detectable. They are absolutely unlike any other kind of star. They are radiating in a completely different way and it has to be electrical. So they are much more like some kind of a radio transmitter than a normal star. And we have to ask – how can a star emit radio wavelengths? Well we do know that stars are magnetic just like the earth. The earth has a magnetic North and South pole. It is the same with stars. Now if we crush them very rapidly the magnetic field inside the star is concentrated, so that neutron stars should be immensely powerful spinning magnets. Now a spinning magnet is an electrical power generator. An experiment any student can do in a physics laboratory is to take a normal bar magnet, spin it around its axis and connect one end of the magnet to a sliding contact on the side. Then a voltage can be measured. It is a rather small one in the laboratory but it is a measurable effect. If we consider the same effect in a neutron star, there would be an enormous voltage. Many thousands of millions of volts between the pole and the equator and this is the key to understanding why Pulsars emit radio waves.

Any neutral atmosphere would be held down to a height of merely a few centimeters because the gravitation pull is so enormous, an extended atmosphere could not exist. However, the electrical forces turn out to be even stronger than gravity and are powerful enough to drag an electrically charged atmosphere into space. The charged particles come from the surface where there is a layer of degenerate matter containing electrons and protons and the atmosphere is called the magnetosphere. Pulsar radiation must be generated somewhere within this region.

Twenty six years after the discovery of Pulsars no one has yet managed to obtain a precise solution to the equations which describe the neutron star magnetosphere. Some parts of it are negatively charged, some parts of it are positively charged and held into the star by the star's own magnetism. The atmosphere rotates with the star within a certain distance, but beyond this it must break away because continuing to rotate would exceed the speed of light, and that is forbidden by Einstein's theory of relativity. The initial distance is usually thousands of miles from the neutron star, but it can be much smaller for the most rapidly spinning Pulsars.

Near the magnetic poles, where charged particles can escape freely from the neutron star, electrons can be accelerated to exceedingly high energy and other processes can then occur. Moving along the curved magnetic field an electron will emit gamma radiation, and gamma radiation generates electron-positron pairs. These newly created particles are, in turn, accelerated, emitting yet more gamma radiation and electron-positron pairs. We get an avalanche effect.

We can get an avalanche effect with the sudden creation of many, many more particles and this effect close to the magnetic poles can produce a flow of electrons and positrons leaving the surface of the star and rushing out into space.

Physicists have been studying electrically charged atmospheres (plasma physics) for many years, especially in relation to producing energy by controlled nuclear fusion, and much has been discovered about radiation from plasmas. One possibility is that regions of positive and negative charges can develop coherent oscillations which generate radiation – something like a laser working at radio wavelengths. Another possibility is that bunches of positive or negative charges are accelerated outwards along the curved magnetic field, again giving rise to coherent radio emission.

The theory that radiation is beamed along the magnetic axis, which is not aligned with the axis of rotation of the neutron star, was first suggested by Dr. Radhakrishnan, who is currently Director of the Raman Institute in Bangalore. This is now generally accepted as being the mechanism which produces the pulsar beam but we still do not know which plasma process actually generates the radiations.

Currently about six hundred pulsars are known to exist and they will be studied very thoroughly. So there is much observational evidence accumulating which should clarify the details still to be determined. But there are complications. For some pulsars the beam consists of several components which could be related to surface features of the neutron star. To think of pulsars as a simple light house producing a single flash is not really quite right. They often produce a complex flash as the beam sweeps across the line of sight to the earth and we have to work out from that what the shape of the beam really is in three dimensions.

In addition to generating beamed radiation neutron stars have other fascinating properties. They are, of course, much more complex than spheres containing neutrons. There is a huge variation of pressure as we go from the surface to the inner regions. We believe that there is an extremely rigid outer crust composed of iron nuclei arranged in a cubic lattice with electrons moving freely between the nuclei. This is degenerate matter of the kind found in white dwarf stars. Further down, as the density and pressure increase, there are more and more neutrons, and below about $2\,km$ from the surface it is mostly neutron matter. In this region neutrons with oppositely directed spin combine in pairs to form composite particles with zero-effective spin, that we call bosons. Remarkably, this stage will be a quantum liquid having no viscosity. In spite of having a temperature around one million degrees, and a density exceeding one hundred million tons per cubic centimetre, it behaves like liquid helium close to absolute zero temperature in the laboratory and becomes a superfluid – having the ability to flow through a capillary tube with no resistance to its motion. As we get really close to the centre and the pressure gets much higher still, there are questions which depend upon physics yet to be discovered. Some of the strange particles we already know, the particles called quarks may exist as a stable, bulk material. But these are speculations. But the general idea of the rigid crust and the liquid interior makes a neutron star look a bit like a hen's egg. It's a shell containing a liquid. Observationally we can test this model due to a discovery made many years ago.

When I discussed neutron stars spinning in space and gradually slowing down that is the first thing that we notice. But looking more closely we occasionally see other effects. Sometimes we observe a change as if the rotation rate of the pulsar has suddenly speeded up. The time interval between the pulses suddenly decreases by a tiny amount, perhaps by only one part in a million but the pulses are so accurate that this can be detected very easily. Now this is very interesting because a well known law of physics says that if you have a spinning body in empty space it's got to spin at the same rate unless it suddenly looses energy. We call this Conservation of Angular Momentum. So if we suddenly see the spin rate increase for an isolated body then that tells us something about the structure of the body. If the body changes shape it can spin faster. Just like a skater. A skater spinning on the spot, with arms outstretched, spins much faster when the arms are suddenly withdrawn. Something like this can happen with the neutron stars. For example if the outside shell was to suffer something like an earthquake which changes its physical shape a little, that effect could actually speed up the outer shell of the star if it can move independently of the liquid interior.

In the course of time the shell must transfer its faster spin to the liquid interior but because this is a superfluid it takes a long time. Measuring these effects confirms that the superfluid interior really does exist, and that the neutrons are behaving as predicted theoretically, which is highly encouraging. Now I have said enough about neutron stars as physical objects. They are immensely fascinating, and understanding them still has problems, but we are on the right track.

Another aspect of neutron stars is that we can use them as accurate clocks to test other laws of physics and one of the areas where this was desperately needed was to check Einstein's more advanced theory of Relativity which is called General Relativity. The ideas of General Relativity are very subtle and have not been absolutely accepted. Recently the Nobel Prize went to Joseph Taylor and his colleague Russel Hulse for the discovery of two neutron stars in orbit about each other. He tells us that this is a wonderful laboratory for carrying out experiments that we can only dream about here on the ground. But nature performs these experiments for us in outer space.

Stars often occur in pairs orbiting about each other and held together by gravitational attraction. Binary stars are quite common objects and I think it is true to say that more than half the stars in the sky are in binary combinations of some sort. So neutron stars can form in pairs and this is what was found in 1975 by Taylor and his colleague.

To get a physical picture of this we can imagine two neutron stars in an orbit which is so tight that the orbit would actually fit inside the sun itself. One of the neutron stars is a pulsar and the presence of the other one can be deduced from accurate timing of the pulses. According to the theory of General Relativity a binary combination like this will emit waves in space time which we call gravity waves. This was a possibility predicted by Einstein but we could never detect it in the laboratory because the effect is far far too weak. The waves of gravitational force – which is essentially what they are – must carry energy away from the system so that the system looses energy and the orbiting neutron stars must come closer together.

As I said earlier, pulsars keep time to better than a millionth of a second and using them as accurate clocks we can detect the shrinking of this orbit. Careful measurements by Taylor and his colleagues over many years confirm that gravitational waves must exist, although they cannot be detected directly as gravity waves. That is a great confirmation of Einstein's theory of General Relativity.

Another effect predicted by Einstein was precession of the orbit of the planet Mercury. According to Newtonian theory the orbit, should remain elliptical and constant in time. But if we put in General Relativity the orbit actually begins to precess around and this is observed for Mercury, but the effect is very tiny. In the case of the binary pulsar the orbital precession, is enormous. What takes over a thousand years for Mercury occurs in two or three days for the binary pulsar – where the axis of the orbital ellipse rotates through several degrees in one year. The effect is many thousands of times bigger.

These phenomena can also be used to measure the masses of these pulsars. It has been calculated that the neutron stars are about one and a half times the mass of the sun, which is a very good confirmation of the predicted mass of a neutron star. Somewhat heavier stars would collapse and become black holes but that's another story, I have no time to discuss.

When space time bends or curves other strange things happen. For example if we look at a light wave passing a star then the light wave is deflected and this is a measurable effect. When space is bent it takes longer for a light wave to pass through it. Once again we find a wonderful correlation between the expectations of Einstein's theory and what is actually observed. There are occasions when radiation from the pulsar passes quite close to its binary companion and this bending of space time is very clearly detectable. This shows some of the excitement in physics which has come from the discovery of pulsars and I would like to conclude by mentioning just one or two final things.

One of the fascinations of studying pulsars is that every ten years or so there is a new discovery which adds up quite new features and raises new questions and generally speaking is of great interest for scientists. And one of these was the 1985 discovery of pulsars which were spinning over a thousand rotations per second. It is hard to imagine a body as massive as the sun spinning so fast. It is a very remarkable situation. We call these millisecond pulsars and they are probably the most accurate clocks in existence.

There is the possibility of using this accurate time to make cosmological observations because there are other objects in space, not neutron stars, which are also emitting gravitational waves. We have heard of black holes. Black holes are closed-off regions of space time which is another prediction of General Relativity. If such regions of space do exist we would not be able to see them because they cannot emit light. But we might be able to detect gravitational waves released by them when they collapse or combine with other objects. There is new physics to be learnt, and astronomy too, by careful observation of these particular pulsars which make such accurate clocks.

But why do millisecond Pulsars spin so rapidly? In binary star combinations a neutron star can start sucking gas from the atmosphere of its companion. If that happens then, because the whole system is spinning, the neutron star acquires some of the spin of its companion and therefore spins faster. The process takes a million years or so but this is how binary neutron stars can become millisecond Pulsars.

I hope I have given in this talk some idea of the wide range of physics, and the excitement, that Pulsars provide and it seems that every ten years or so something new occurs. They have definitely opened up a new chapter in astronomy and physics and I think I was extremely fortunate to have begun this story back in 1967.

Thank you very much.

Is the Future Given? Changes
in Our Description of Nature

Ilya Prigogine

ULB, Bruxelles, Belgium

Fig. 1. Ilya Prigogine delivering the B.M. Birla Memorial Lecture

Prof. Ilya Prigogine was born on the 25th of January 1917 into the family of
a Chemical Engineer of the Moscow Polytechnic. Those were the tumultous
years of the Russian Revolution and the Prigogine family left for Germany in
1921. After a few years in Germany they settled down in Belgium in 1929.

Ilya Prigogine attended Secondary School and University, in Belgium,
studying Chemistry at the Universite Libre de Bruxelles. He was also very
interested in History, Archaeology and Music. Infact he was an accomplished
piano player. He was also deeply interested in Philosophy, particularly Western
Philosophy. His critique of some of the philosophers in the light of modern
physics can be found in many of his books.

At Brussels, Prigogine developed a School for the study of Thermodynamic
Principles applied to several disciplines, including Biology, Chemistry, Physics,
Sociology and so on. His pioneering work was in studying Thermodynamics
far from the equilibrium. This lead to mathematical models of dissipative

B.G. Sidharth (ed.), *A Century of Ideas*, 65–75.
© Springer Science+Business Media B.V. 2008

systems and self organization, something which seemed to be contrary to the usual Thermodynamic drift towards total disorder. He was awarded the 1977 Nobel Prize in Chemistry for this work. Numerous honors and awards were also heaped on him over the years.

For several decades Prof. Prigogine served as Professor of Physical Chemistry and Theoretical Physics at the Free University of Brussels. He was also the Director of the Center for Statistical Mechanics at the University of Texas, Austin, USA.

I had enjoyed warm rapport with Prof. Prigogine for many years. He was a keen observer of things around him. When he came to India he spoke at length about what he had encountered. A large number of flights linking different cities. Indians constantly on the move, flying from one place to the other. "The Indian economy is on the move" he said. His comment on the somewhat chaotic traffic in India was that Indian drivers are far more interactive than those in the West. He also noted with approval that India had a very early start – even in the very early hours, people were out for their morning exercise. So also he noted that the cities were awake till late – unlike Brussels he added.

We had several discussions on philosophical aspects, including the Western, the ancient Indian (Upanishadic) and Buddhist perspectives as also on matters of physics. He seemed to like Buddhist thought, compared to the others. On one occasion he said, "Marshak told me that at a meeting he had mentioned the work of one of his students (E.C.G. Sudarshan). 'I threw away his Nobel Prize', Marshak said."

He had a keen interest in antiques and was a collector of several pre Columbian artefacts as also some from India. In fact on his visit he even took some of these with him, including an unusual depiction of Nataraja – this was of course an expensive handicraft, not an antique. He was also an avid shopper. Back home, he would proudly show his collection to visitors and venture explanations.

Prof. Prigogine always evinced keen interest in my work and would make me explain some details to him. He wrote to me as late as 2003, "I agree with you that space time has a stochastic underpinning". I was looking forward to further discussion. But a few days later I received an email from his Secretary that he was no more.

Summary. According to the classical point of view, nature would be an automaton. However, today we discover everywhere instabilities, bifurcations, evolution. This demands a different formulation of the laws of nature to include probability and time symmetry breaking. We have shown that the difficulties in the classical formulation come from too narrow a point of view concerning the fundamental laws of dynamics (classical or quantum). The classical model has been a model of integrable systems (in the sense of Poincare). It is this model, which leads to determinism and time reversibility. We have shown that when we leave this model and consider a class of non-integrable systems, the difficulties are overcome. We show that our approach unifies dynamics, thermodynamics and probability theory.

1 Introduction

I feel very moved by the kindness shown to me. I don't know if I deserve so many honors. I remember that some years ago a Japanese journalist asked a group of visitors why they are interested in science. My answer was that I feel that science is an important way to understand the nature in which we are living and therefore also our position in this nature. I always felt that there are some difficulties in the descriptions of nature you find currently. I would quote three features. First of all, nature leads to unexpected complexity. This is true on all levels. It is true in the case of the elementary particles; it is true for living systems and, of course, for our brain. The second difficulty is that the classical view does not correspond to the historical time-oriented evolution, which we see everywhere around us. The universe is evolving. That is the main result of modern cosmology with the Big Bang. Everywhere we see narrative stages. They are events in nature. An event is something, which may or not happen. For example, the position of the moon in one million years is not an event as you can predict it, but the existence of millions of insects as we observe is an evidence of what we could call creativity of nature. It is indeed difficult to imagine that the information necessary existed already in some way in the early stages of the universe.

These difficulties have led me to look for a different formulation. This problem is a continuation of the famous controversy between Parmenides and Heraclitus. Parmenides insisted that there is nothing new, that everything was there and will be ever there. This statement is paradoxical because the situation changed before and after he wrote his famous poem. On the other hand, Heraclitus insisted on change. In a sense, after Newton's dynamics, it seemed that Parmenides was right, because Newton's theory is a deterministic theory and time is reversible. Therefore nothing new can appear. On the other hand, philosophers were divided. Many great philosophers shared the views of Parmenides. But since the nineteenth century, since Hegel, Bergson, Heidegger, philosophy took a different point of view. Time is our existential dimension.

I want to show you that the dilemma between Heraclitus and Parmenides can now be put on an exact mathematical framework. As you know, we have inherited from the nineteenth century two different world views. The world view of dynamics, mechanics and the world view of thermodynamics. Both views are pessimistic. From the dynamical point of view, everything occurs in a predetermined way. From the thermodynamic point of view, everything goes to death, the so-called thermal death. Both points of view are not able to describe the features, which I have mentioned before. Matter was generally considered as a kind of ensemble of dust particles moving in a disordered way. Of course, we knew that there are forces. But the forces don't explain the high degree of organization that we find in organisms.

For classical physics including quantum physics, there is no privileged direction of time. Future and past play the same role. However we see an evolutionary universe on all levels of observation. The traditional description is deterministic, even in quantum theory. Indeed, once we know the wave function for one time, we can predict it for an arbitrary future or past. This I felt always to be very difficult to accept. I liked the statement by Bergson: time is "invention".

But the results obtained by classical or quantum mechanics or classical thermodynamics contain certainly a large part of truth. Therefore, the path, which I followed over my whole life, was to show that these descriptions are based on a too restricted form of dynamics. We have to introduce a more general starting point. The first step in this direction was an observation, which I made at the beginning of my PhD, in 1945, that non-equilibrium leads to structure. For example, if you consider a box containing two components, say N_2 and O_2, and you heat it from one side and cool it from the other, you see a difference of concentrations. For example, N_2 may be more concentrated at the hot side. Of course, when you consider the box in thermal equilibrium, the concetrations become uniform. Much later, thanks to the collaboration with Prof. Glansdorff, we found that far from equilibrium there appears what we called dissipative structures. These new structures have become quite popular, everywhere one speaks about non-equilibrium structures, self-organization. These concepts have been applied in many fields including even social sciences or economic sciences. But I could not stop at this point because thermodynamics is macroscopic physics, so perhaps it is the fact that these systems are large and that we have no exact knowledge of their time evolution that would give us the illusion of irreversibility. That is the point of view adopted by most people even today. However, my main interest was to show that the difficulty comes from the fact that dynamics, classical or quantum has to be put on a more general frame.

Let me make here a short excursion into theoretical physics. To describe our nature, we need observables as space and time. You know that Einstein's great idea was to relate space and time to the properties of matter. But I want not to consider relativity, but classical systems, such as a pendulum, planetary motion or the motion of particles in a gas. To describe classical systems of this type, we need two kinds of variables: coordinates q and momenta p. In classical theory, a dynamical system is described by the so-called Hamiltonian H. The Hamiltonian is simply the expression of the energy in terms of the observables p and q. Once we have the Hamiltonian, we can predict the motion through the so-called canonical equations (the dot means the time derivative.)

$$\dot{p} = \frac{\partial H}{\partial q} \qquad \dot{q} = -\frac{\partial H}{\partial p}$$

At the initial time, the observables are q_0, p_0. Time going on, they change into $p(t), q(t)$. The observables q, p are called the 'canonical variables'. Now, a very important point is that there are various choices of canonical variables q and p. This is studied in the basic chapters of classical physics. It is natural to choose the set of variables q, p, such that the solutions of the canonical equations of motion are as simple as possible. It is therefore natural to try to choose them in such a way that we eliminate the potential energy. The Hamiltonian then depends only on p. We have then $H(p)$ and $\dot{p} = 0$. Momenta are constant and the time derivative of the momenta vanishes.

For a long tie it was considered that this was always possible. We could always eliminate the coordinates in the Hamiltonian. But Poincare, at the end of the nineteenth century, made a fundamental discovery. He discovered that this elimination was only possible for a class of dynamical systems, which he called "integrable systems". For example, in a gas, with many particles, this transformation would correspond to going to a representation in which each particle moves independently. When this is possible, the momenta are also called the action variables J and the coordinates α, the angle variables. I have to be a little more specific. Consider a system in which the Hamiltonian has two parts

$$H(J, \alpha) = H_0(J) + \lambda V(J, \alpha)$$

We have then one part, H_0, which depends only on momenta (the action variables) but there is also a perturbation λV depending on both J and α. λ is a parameter measuring the intensity of the perturbation. By definition, for H_0, we know the action variables. Then for H including λV, we ask if we can construct new action variables, J', which would depend analytically on the old ones. That means that the Hamiltonian H can be written $H(J')$ with

$$J' = J + \lambda J^{(1)} + \lambda^2 J^{(2)} + \cdots$$

What is the meaning of action variables? They represent independent objects, as interactions are eliminated or better to say included in the definition of these objects. This transformation theory has been intensively studied in the nineteenth and twentieth centuries. We can in general introduce new momenta and new coordinates related to p and q by $p' = U^{-1}p, q' = U^{-1}q$, where U is a so-called unitary operator. These transformations are made in such a way that the Hamiltonian equations remain valid. U plays an essential role both in classical and quantum mechanics. An important property is this distributivity of U. That means that U acting on a product is equal to the product of the transformations. $U^{-1}(AB) = (U^{-1}A)(U^{-1}B)$. There are other remarkable properties of unitary transformations here but there is no place to go further into this.

It is remarkable that orthodox quantum mechanics used as a model classical integrable dynamical systems. The basic difference is that the observables are now no longer numbers but operators. There are again various representations of the operators related by unitary transformations. Let us only remind that, according to every book on quantum mechanics, in the representation in which q is a number, p is the operator $i\frac{\partial}{\partial q}$ and we have the commutation relation $qp - pq = \frac{\hbar}{i}$. This is the basis of the Heisenberg uncertainty relations. For non-integrable systems, the situation, as we shall see now, is quite different.

2 Non-integrable Systems

After this short introduction to integrable systems, we go now to non-integrable systems. There are of course many classes of non-integrable systems, that is of systems for which there exists no unitary transformation, which eliminates interactions. We shall consider a specific class of non-integrable systems. That is the class where there exist resonances. What is a resonance? Consider a particle, like a harmonic oscillator, in a field like in electromagnetism. Suppose that the particle frequency is ω_p while the field forms a continuous set of frequencies starting from 0

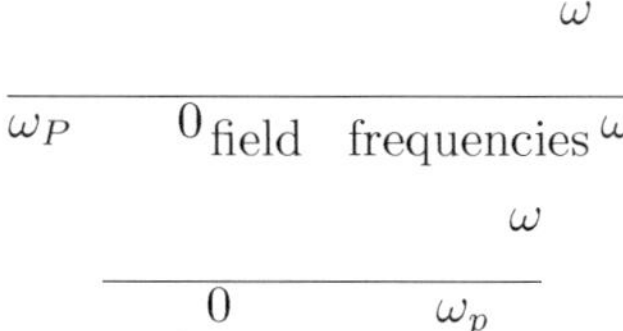

Then there are two situations, either the frequency of the oscillator ω_p is below all the frequencies of the field or the frequency of the oscillator is somewhere in the domain of the frequencies of the field. These are two very different situations. If the frequency of the oscillator is outside the field, nothing special happens. But if it is inside, we have a so-called excited state and this excited state decays by emitting a photon to a ground state.

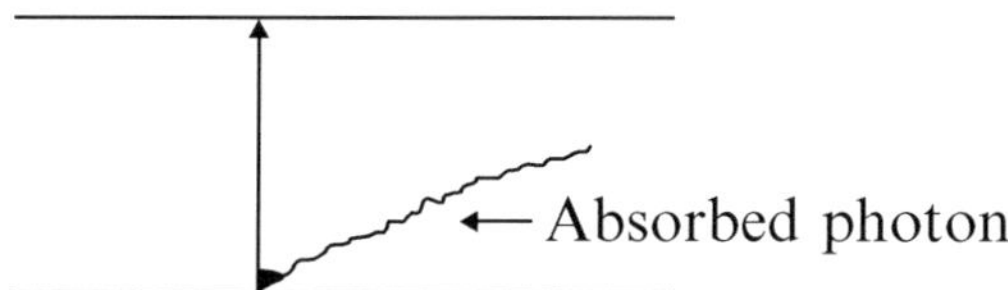

This is the well known Einstein and Bohr mechanism for the description of spectral lines. It is generally expressed by saying that the particle is dissolved in the continuum. We have a de-excitation process. There exists of course also an excitation process when the photon falls on the ground state.

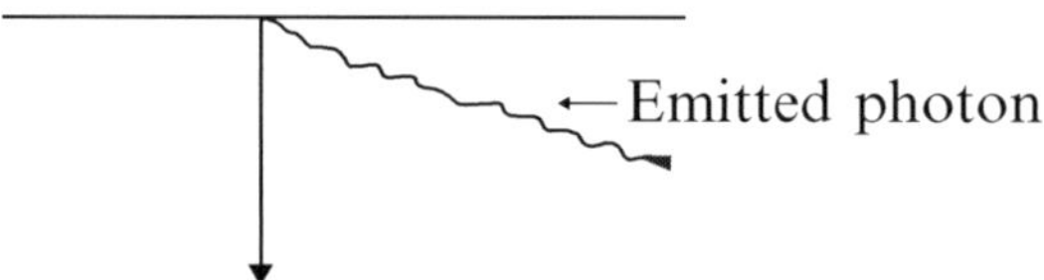

The interactions between the field and the oscillators are described by resonances. The fundamental result of Poincare was to show that such resonances lead to difficulties through the appearance of divergent denominators.

An example is

$$\frac{1}{\omega_k - \omega_l}$$

$$\text{particle} \rightarrow \qquad \leftarrow \text{field}$$

This difficulty was known already by Laplace. How to overcome this difficulty? We have shown that the resonances can be avoided by suitable "analytic continuation"; that means that one has to put small quantities in the denominator to avoid the infinities. Of course, there are some specific mathematical problems to be overcome here but it can be studied in the original papers [1]–[10].

In short, our basic idea was therefore to eliminate the Poincare divergences and to extend the idea of unitary transformations. Instead of the formula we have already written for unitary transformation, $q' = U^{-1}q$, $p' = U^{-1}p$, we now obtain $q' = \Lambda^{-1}q$, $p' = \Lambda^{-1}p$.

The unitary operator U has been replaced by the operator Λ (which is a star-unitary operator but that doesn't matter here). We have an extension of canonical transformations. In other words, we have now a new representation of observables and an extension of the classical theory. Even in classical theory, it is very important to choose the right representation. For example, if you consider a crystal with vibrating atoms you can go to a representation in which you have normal coordinates that means independent motions and then you can define the basic frequencies. Similarly here by using the new representation, you can come to expressions of motions, classical or quantum, in which there appear quantities such as transport quantities, reaction rates, approach to equilibrium.

Now the Λ, which replaces U, has very interesting new properties. First of all, it is a non-local transformation. In other words, classically people were thinking in terms of points but here we have to speak in terms of ensembles. We cannot any more make a physics of points but we have to make a physics of distributions. This means that we have a statistical description. That also means that we have to give up classical determinism.

The second fundamental property of Λ is that we have no more distributivity. More precisely we have $\Lambda^{-1}AB \neq \Lambda^{-1}A \cdot \Lambda^{-1}B$. This opens a whole new domain of classical and quantum physics. We have the appearance of new fluctuations and new uncertainty relations. For example, the Λ operator acting on a product of coordinates is not the product of the transformed coordinates. There is an uncertainty in position. Let me give an example. In statistical physics, an important role is played by the so-called Langevin equation, where γ is the friction, and noise:

$$dp_1(t)/dt = -\gamma p_1(t) - m\omega_1^2 x_1(t) + B(t)$$
$$dx_1(t)/dt = -\gamma x_1(t) + p_1(t)/m + A(t)$$

These equations describe the damped harmonic oscillator with random momentum. This corresponds, for example, to the motion of a heavy particle in a thermal medium and it is one of the most important results of statistical physics.

Now recently S. Kim and G. Ordonez have shown that using our new transformation Λ, you derive exactly the Langevin equations and therefore also the basic properties studied in statistical mechanics. The Langevin equation has a broken time symmetry. This is not due to approximation but expresses that $x(t)$ and $p(t)$ are Λ transforms. The Langevin equation corresponds to a system in which resonances between the Brownian particle and the thermal medium play an essential role. We have also obtained the quantum Langevin equation. The operators are again Λ transforms. Uncertainty relations can now be established for x and p separately. The whole space-time structure is altered. These are fundamental results. Dynamics and probability theory were always considered as separate domains. In other words, statistical theory, noise, kinetic equations were considered as coming from approximations introduced into dynamics, classical or quantum. What we show now here is that these properties, noise and stochasticity are directly derived from a more general formulation of dynamics. These are the consequences of non-integrability while integrable systems, which were used as a model for classical and quantum physics, refer in fact only to exceptional systems. We are living in a nature in which the rule is non-integrability. And in non-integrable systems we have quite new properties. The new properties are: first of all, the appearance of new fluctuations, therefore no more determinism, the appearance of a privileged direction of time that is due to the analytic continuation and non-distributivity leading to new uncertainty relations, even in classical physics.

These new properties come from the fact that what we use is analytic continuation and also that the analytic continuation of a product is not the product of the analytic continuations. When we observe the Langevin equation, the coordinate x and the momentum p have to be understood as non-unitary transforms of the initial variables. And the new transforms lead to stochasticity and probability. In the classical point of view, we may either start from an individual description or with ensembles. Gibbs and Einstein have shown that thermodynamics is based on the theory of ensembles. This, as we have already mentioned, was considered as the result of approximations ("coarse graining"). This is no more so for our class of non-integrable systems. The ensembles point of view is a consequence of the Λ transformation. Λ transforms a phase point into an ensemble. More precisely, the Liouville equation is transformed into a kinetic equation. This, I believe, closes a controversy, which goes back to Boltzmann (1872).

3 Irreversibility

We want now to go to a different aspect. This aspect is related to a different description of elementary processes, unstable particles or quantum transitions. In a sense, it is a very happy circumstance that these systems are non-integrable. If you could, in the examples of the interaction between the oscillators and the field, apply a unitary transformation, you would not be able to observe the quantum transitions from one level to the others. Electrons, photons are only observable because they interact and participate in irreversible processes. The basic idea of unitary transformation of integrable systems is that you could, in one way or another, eliminate interactions. But interactions are a fundamental part of nature which we observe and, in non-integrable systems, interactions can not be eliminated. Think about a gas. In a gas, even if it is in equilibrium, collisions continue to occur and interactions are never eliminated. Collisions give rise to thermal motion. There are limits to reductionism. We have applied our method to a number of problems such as unstable particles or radiation damping (details can be found in the original publication – [8]–[11].

Once we have irreversibility it is clear that we have also some form of the second law of thermodynamics, that means entropy.

Boltzmann had the ambition to become the Darwin of physics. He studied the collisions in dilute systems and showed that you can find a function, which plays the role of entropy. This led to a lot of controversies. Poincare wrote that there was a basic contradiction: on one side, to use classical mechanics; on the other hand, to come out with entropy which is time oriented. We can now understand what was the reason. Boltzmann tried to apply classical mechanics to non-integrable systems. Gas cannot be an integrable system because then it would never go to equilibrium. For example, all momenta would be invariants of motions. So we need non-integrable systems. And once we have non-integrable systems, then Boltzmann's equations are exact consequences of the extended dynamics.

Indeed, we have shown, together with Tomio Petrosky, Gonzalo Ordonez, Evgueni Karpov and others that we can formulate the second law in terms of dynamical processes. There were always two points of views. The point of view of Boltzmann, stating that the second law is probabilistic and comes ultimately from our ignorance and the point of view of Planck that the second law, the entropy production is a consequence of dynamics. Consider the problem of resonances, which I described a little earlier, we have shown that the decay of the excited state with the emission of the photon is an irreversible process leading to entropy production. This is not astonishing because, in a sense, an excited state contains "more energy" than the ground state. This supplementary energy can then be distributed on all the degrees of freedom of the field. And we have shown that the inverse process is also possible; that to bring an atom into an excited state, we need a process, which brings negative entropy to the atom, which is then used to excite it. In a sense, our whole vision of the universe around us is an example of non-equilibrium systems. We have particles, with mass, and we have photons, without proper mass. Particles with mass should, from the thermodynamical point of view, dissolve into a continuum. Probably the main event in the history of our universe, in the Big Bang is this differentiation. We have massive particles floating in a bath of zero mass objects like the photons.

4 Conclusions

We come to a different concept of reality. Laplace and Einstein believed that man is a machine within the cosmic machine. Spinoza said that we are all machines but don't know it. This does not seem very satisfactory. However, to describe our evolutionary universe, we have only taken very preliminary steps. Science and physics are far from being completed, as some theoretical physicists want us to believe. On the contrary, I think that the various concepts, which I have tried to describe in my lecture, show that we are only at the beginning. We don't know what exactly corresponded to the Big Bang, we don't know what determines the families of particles, we don't know how the biological evolution is evolving.

May I finish my lecture with some general remarks. Non-equilibrium physics has given us a better understanding of the mechanism of the emergence of events. Events are associated with bifurcations. The future is not given. Especially in this time of globalization and the network revolution, behavior at the individual level is the key factor in shaping the evolution of the entire human species, just as a few particles can alter the macroscopic organization in nature, show the appearance of different dissipative structures. The role of individuals is more important than ever. This leads us to believe that some of our conclusions remain valid in human societies.

A famous saying of Einstein is that time is an "illusion". Einstein was right for integrable systems but the world around us is basically formed by non-integrable systems. Time is our existential dimensions. The results described in this paper show that the conflict between Parmenide and Heraclitus can be taken out from its metaphysical context and formulated in terms of modern theory of dynamical systems.

Thank you very much.

Acknowledgements

The results described in this paper are the fruit of the work of the Brussels-Austin group. I shall not thank individually each member of this group but I want to make an exception for Prof. Ioannis Antoniou, Dr. Gonzalo Ordonez and Prof. Tomio Petrosky. It is their work, which has led to the formulation of the extension of canonical transformations. I acknowledge the European Commission Grant Nos. HPHA-CT-2000-00015 and HPHA-CT-2001-40002, the Engineering Research Program of the Office of basic Energy Sciences at the US Department of Energy, Grant No. F-0365, the National Lottery of Belgium, and the Communaute Francaise de Belgique.

References

1. I. Prigogine *Non-Equilibrium Statistical Mechanics*, Wiley, New York (1962).
2. I. Prigogine *From Being to Becoming*, Freeman, New York (1980).

3. T. Petrosky, I. Prigogine and S. Tasaki, "Quantum theory of nonintegrable systems", *Physica* A **173**, 175–242 (1991).

4. I. Antoniou and S. Tasaki, "Generalized spectral decomposition of mixing dynamical systems", *Int. J. Quantum Chem.* **46**, 425–474 (1993).

5. I. Antoniou and I. Prigogine, "Intrinsic irreversibility and integrability of dynamics", *Physica* A **192**, 443–464 (1993).

6. T. Petrosky and I. Prigogine, "Poincare resonances and the extension of classical dynamics", *Chaos, Solitons and Fractals* **7**, 441–497 (1996).

7. T. Petrosky and I. Prigogine, "The Liouville space extension of quantum mechanics", *Adv. Chem. Phys.* **99**, 1–120, ed. I. Prigogine and S. Rice, Wiley (1997).

8. G. Ordonez, T. Petrosky and I. Prigogine, "Space-time formulation of quantum transitions", *Phys. Rev.* A **63**, 052106 (2001).

9. T. Petrosky, G. Ordonez and I. Prigogine, "Space-time formulation of quantum transitions", *Phys. Rev.* A **64**, 062101 (2001).

10. E. Karpov, G. Ordonez, T. Petrosky and I. Prigogine, "Quantum transitions in interacting fields", *Phys. Rev.* A **66**, 012109 (2002).

11. I. Prigogine, S. Kim, G. Ordonez and T. Petrosky, *Stochasticity and time symmetry breaking in Hamiltonian dynamics*, submitted to Proc. Solvay conference in Delphi, Greece (2001).

12. G. Ordonez, T. Petrosky and I. Prigogine, *Microscopic entropy flow and entropy production in resonance scattering*, submitted to Proc. Solvay conference in Delphi, Greece (2001).

13. E. Karpov, G. Ordonez, T. Petrosky and I. Prigogine, *Microscopic Entropy and Nonlocality*, Proc. Workshop on Quantum Physics and Communication (QPC 2002) Dubna, Russia (2002), Particles and Nuclei, Letters. No. 1 [116], 8–15 (2003).

Bubbles, Foams and Other Fragile Objects

P.G. de Gennes

College de France, Paris, France

Fig. 1. Pierre-Gilles de Gennes with Mr. H.K. Kejriwal, Vice President of B.M. Birla Science Centre

Pierre-Gilles de Gennes was one of those rare scholars who was also a very enthusiastic popularizer. Born in Paris in 1932 he majored from the Ecole Normale in 1955. For the next four years he was a research engineer at the Atomic Energy Centre in Saclay. His work pertained to neutron scattering and magnetism. From 1959 he was a Post Doctoral fellow at Berkeley with Prof. C. Kittel and subsequently served the French Navy for over two years. In 1961 he took up an Assistant Professor's post in Orsay where he started a group on superconductors. In 1968 he started work on liquid crystals, becoming a professor at the College de France in 1971. Here he participated in a joint group on polymer physics. From 1980 his interest in interface problems began. He worked on dynamics of wetting and the chemistry of adhesion.

Prof. De Gennes got the Nobel Prize in Physics in 1991. This apart he received the Holweck Prize from the joint French British Physical Society, the Ampere Prize of the French Academy of Science, the Gold Medal from

B.G. Sidharth (ed.), *A Century of Ideas*, 77–85. 77

the French CNRS, the Matteuci Medal of the Italian Academy and numerous other distinctions including the Life Time Achievement in Science Award of the B.M. Birla Science Centre. He was a Member of the French Academy of Sciences, the Royal Society, the American Academy of Arts and Sciences, the National Academy of Sciences, USA, amongst other distinctions.

Prof. De Gennes enjoyed delivering both scholarly and popular lectures which are invigorated by his great sense of humour.

He died in Orsay on 18 May, 2007.

First of all I am immensely pleased to be here. I have had great interest in India over many years. I want to remember that we are standing in the Science Museum and the aim of this Science Museum is to make science stimulating and attractive to young minds. Also to older minds.

Bubbles have inspired people for centuries. I think this is an interesting subject. I think also that education in recent times will be ultimately useful and I hope to show this to you at the end of this talk. Let me start by saying what a bubble really is. It is a thin sheet of water and to this water we must add a little soap. That is very critical and I will try to give you a feeling why we need the soap. These things have color, these things are mobile. Not only the whole globule will move around but if you take your John Boy and you have him look at it closely, you will notice that on the surface, there are all strange movements forming and deforming and ultimately it has a short life. So, where is the life for the bubble, where is color for the bubble – I want to talk about all of these.

First thing the color: Let us start with a man of importance, Isaac Newton. You have quoted statements according to which I have some resemblance to Isaac Newton. My reaction to this is when I speak to my students I am something like a nano-Newton, which means a brilliant spark of Newton. But I say this not only because of modesty but also because a nano-Newton is exactly the force which was needed to separate two atoms in a molecule. And once we know this, we have learnt something. This is why you have Science Museums too. Well, Newton in his early youth had an understanding of the real nature of light, related to experiments such as soap bubbles. He realized that light propagates by waves. Just as, if you stand by the beach you get successive waves coming and if they come along a very steep rock, you see the waves coming out again, this is exactly what happens on the two sides of this thin water sheet which makes the wall of the bubble and if the distance between waves is just equal to the thickness of this region. However, the amplitudes of the two waves superpose. And in this way you get something which gives you a colored frame because the wavelength is different for blue light and for the red light. Blue light has waves which are very close to one another, red light has much more spread out light waves. So, the film will interact with blue light when it is thin and will interact with red light when it is thick, and this way it gets colored. By seeing the colors you can see the differences in the thicknesses of various portions.

Now, let me embark on a slightly more delicate question, why do we have to put soap? Let me start with some experiments. Suppose I have a brush and I dip it into water. Very often this is a painting brush. In the dry state it will have different hairs pointing in different directions. As soon as I get my brush wet, it coalesces to a single unit. What makes it so? Well, being a physicist I will describe it in terms of a brush which has only two hairs. If I take this object and dip it in water, after getting it out of the water, I find that the two hairs are linked by a film of water. And this film is not happy. The skin of the water is the region of unhappiness. Let's say that the molecules which are near the skin see less neighbors than the molecules which are inside, so we claim that they are unhappy. And for that reason, there is a tendency to make this skin thinner. All liquids have this property. They want to hide their skin. We will always manage to cope with the situation by exhibiting the minimum amount of skin.

We now come to Thomas Young. He was a doctor by trade. But he was interested in physics. And in fact we know his name when we study physics for three different reasons. The first reason is related to elasticity when we look at how we can deform a beam of iron for instance, we use a quantity which is known as Young's modulus. In an other direction, Young was interested in light, the same sort of wave aspects which Newton had studied and he invented the beautiful experiment which is called Young's Pin Hole Experiment which is an interference experiment proving the wave nature of light in a beautiful way. This is the second context.

Thirdly, Young was interested in duplex and films and things of this sort. He was the first to fully appreciate the consequences of this remarkable feature that a liquid like water wants to show as little skin as possible. Apart from that Young had a few hobbies. The most significant of his hobbies was he wanted to decipher the Egyptian heiroglyphies and he was not an amateur. He was in competition with French scholars, another example of Anglo-French competition. This was in 1805 or so, and Young was beaten by six months. But even if he was defeated we should remember people like this, who were able to do a lot in physics and at the same time to do archealogy at the professional level. My impression is that our generations are very far from achieving things of this sort.

Let me return to my films now. I will possibly show you what this problem is of showing as little skin as you can. And this is very much like a glamorous young lady who wants to pretend that she is very cute. But you can constrain a film much more than you can constrain a young lady. Here is the first example. If I take a ring, may be a piece of wire, dip this piece of wire in soapy water, if I take it out, well, the film manages to be flattened. That is the form of the minimal surface acceptable for this film when it is forced to touch the wire. If I take two wires, it is already less obvious, suppose I arrange two wires and arrange to keep a film in this region, what will be the shape? This shape will be a little complex. You see the shape is a little bit weakened. If you want to show little skin, do it better by being thinner. A very important notion for ladies.

Let me show you a further example which is very beautiful. Take a guiding wire, which has the shape of a screw, and the film is lying on the screw and you see the film takes a very complicated shape. Mathematicians love these objects, we call them minimal surfaces, meaning minimum in exposed skin. And we have worked a lot on it.

Let me come to an experimentalist who was the first to study these things. His name was Plato. He was not French. He was from Belgium. A very staunch and a very great man doing beautiful experiments and in a very impressive way. There is a part in the note books of Plato where he recalled his experiments. He says: "From the moment on, the experiments which I will describe will not be performed by myself, but by my assistant Mr. X". What is hidden in this very discrete sentence is the fact that Plato was becoming blind because he had looked at the sun too long for certain of his experiments. That wisdom had not stopped him from working. Well, Plato studied many of these strange surfaces, and really contributed a lot to the science.

Now let me come to, surface tension as we call it. Here is an experiment which children in France should do in schools every day. Children should be asked to wash the dishes after dinner. Essentially no French child does this experiment any more for various reasons. But I regret it very much. If you do it, you put water in some tank and you pour a very small amount of a magic product, which we call a surfactant or a detergent, and immediately you get foam. Now what has happened? The molecules which make this magic product are very small molecules of the size, now called nanometer, a billion per meter. But they are very strange molecules. They are made up of two parts. These two parts have a very different affinity. There is one part which we call the head, which likes water, and there is another part which we call a tail, which hates water. Again to my friend students I very often say that it is very similar to what we see when we observe married couples. Let us take an example which appears relatively innocent. If we observe a couple watching television – the man will want to watch Football and the lady would like to watch some romantic story in the style of Santa Barbara or something like that. Well, how do we solve the problem – even for a single couple or for an assembly of couples, it is very interesting. Here we are talking about the single couple problem. Suppose I am washing the dishes or preparing a batter and putting a few of these molecules into water, then the head is happy and the tail is unhappy. That will move around for sometime and ultimately it will find the surface of the water. And it will discover that it is much happier there because the head cannot stay in its lovable environment which the tail escaped from. Now saying that this molecule is happy here, translates in our physics language to saying that the energies are lower, the system is indeed more stable. That means that we have reduced the surface tension of water. So, if we have less energy to spend in showing skin, then indeed we can show a lot of skin, this is what we have if we have a foam. So, this experiment is really related to this remarkable reactant property of surfactant molecules to go to its surface. This one brings me to a beautiful experiment.

Sometimes I ask my students how would you study a film of soap which is now so thin? This is just a little decoration on one side. And it is nano-meters instead of being micrometers. A very minute object. How would you study it? My students react very fast and say this is very small and we must use some sort of gradation which has a wavelength comparable to the size – one is X-rays, wavelets of X-rays are very much in this category. Another is the neutron. So, for instance, we go to a neutron reactor and so on. But I tell them, that you are taking a very big hammer to break a very small nut. The reason is that Benjamin Franklin did it much better. Franklin was at that time living in London. He had a vast education. He knew that if you pour oil on the sea, you, to some extent quieten the waves. Well, he decided to make this statement more precise. So, he went to Kleper Comma. He had a very nice pond and Franklin chose this, when there was a little wind so that there was a little wavelet on the surface, very easy to observe these wavelets. Franklin had brought a little vial of Oleic oil. Oleic is a very standard oil. He took a spoonful of that and he poured it on the pond. He noticed immediately that all around this point, there were no waves left. What really amazed him is that this was a huge effect, spreading over a yard or so, for example. You can go to the pond and see how large it is. It is much larger than this room. And a single spoonful did it all. He noticed this. He wrote it down and unfortunately, I am going to say, he missed his Nobel prize.

He missed one part in the reasoning. The fact is that he knew the volume of oil he had put. He knew the area on which it was spread. Now, we have been taught in school that volume is the product of area by height. So, if he had just divided his volume – spoonful, by this area – which was may be 500 sq yards or something like that, he would have measured the size of molecules. He had all the numbers required to measure the size of molecules.

Unfortunately, he did not do it himself, it was Lord Halley a hundred years later who really did it. But this was a beautiful experiment and this is a sort of experiment that I want my students to memorize. Much more than that of the neutron source experiment!! Because it is what we need in real life. Whenever you begin to attack a problem, here measuring the size of a molecule, before embarking in this enormous manner you should always try and think in a simple way to do it. You know various reasons for that. One reason is with respect to the tax payer. I always say that we scientists should not forget that some tax payers are paying for our research. And I think we should keep this always in mind. Well, from the point of view of a tax payer it is a great danger to have science with bigger and bigger tools. And in many cases it is not the first thing to do. Sometimes it may be necessary later but it is not the first thing to do.

Another aspect is related to industry. You asked whether I have interacted with industry? Well, suppose you are interested in one of these things, let me take a concrete example – you make photographic film. Now this is a very sophisticated object. There are something like sixteen layers on top of each other which are sensitive to different colors and so on and you want to sensitively dip your film into a certain trough and take it out. Now instances like this have all sorts of problems, there are surfactants and there are molecules which are necessary for the process where evaporation is quite delicate. If something gets unhappy in the operation and you are loosing color film, by a kilogram every minute, you better think of a simple way of understanding the problem. You would not face the problem by bringing your machine to a neutron facility. You will solve it by having an intelligent idea with minimal means.

Franklin's spirit is as important now as it was in the old days. And we should never forget it. Let me mention a lady, who is an important item in this story. This is Aganise. She was a normal lady. Don't expect Marylin Manroe. But she was a great lady. She had essentially no science education. But her brother, being a male, had science education, but not her. She was physically working in her kitchen. She had observed that when you have surfs coming out from the soap, sometimes if you want to dispose these surfs a simple thing to do, is to scoop it. To do this, she brought in something which was very profound. The problem was that in the early experiments on surface tension we were always working with polluted water. And remember in this washing experiment a very small amount of the clever polluter is enough to change the surface properties completely. And people had not been able to dispose of that. Now, she came and she said the way to purify this is precisely to get rid of this surface layer which is impure, by scooping. Doing this repeatedly, she got the most beautiful measurements of surface tensions. She wrote to Logo Rali. He got very excited and she became famous. And she was the first lady scientist for many generations. Great woman.

I am still at the birth of my bubble.

Let me tell you, how a bubble is born. And the experiment here would be closer to something like beating eggs. Suppose I have a piece of my beater, which is a wire, which is moving up, and pulling from the bottom of eggs in water, a film. But, why on earth my egg beater has moved from one place to another. Why is this film stable? Because, there is some surfactant on the surface. It was plated and it was surfaced with a coercant liquid. Then I pulled out fast, this surfactant did not follow very well. There was not much of it available totally so that there is less surfactant. And this means that if you notice surface tensions are higher here and there, its like paving your way through the crowd. You take a big crowd on a sunny day, we are all pushing each other with our elbows. And regions where you have a dense crowd there is high pressure. Well, this is the same here. At one place you have high surface pressure, elsewhere you have a low surface pressure. And the difference between the two is the frame which counteracts the wind. This is the more precise reason why you need soap. If you did not have the soap, weight would bring the film back immediately to its normal position. This would be like building a tower without walls.

So, this was the bubble.

Now, let's go rapidly to the agent. One thing is that when you prepare a film, a beautiful way of doing it is in a horse shoe frame shape, and to pull it out from the surf of soapy water. If you watch such a film, the film originally is colored everywhere, but after sometime, it becomes black on the top region. In fact this means that it has become so thin that the two waves which hit the two sides destroy each other completely and you get no refracted lines. This is what we call a black film. Now, let me spend a minute talking about these black films.

Originally the man who studied the black films was Newton. And he realized they represent a very aged version of a film and the film had become very thin. So, these films were called Newton Films. But, in fact, people discovered around forty years ago, that this was not the earliest reference, on black films. The earliest reference is a Babylonian script from something like 1500 BC. I happened to work at a place which has the largest Syrian library in Western Europe. I visited these people and said I want this and they showed me records of this, a huge amount of literature.

The Babylonians were very much like us. They wanted to know the future. And the way to know the future is to use a random process which we understand nothing about, such as drying of the inside of coffee cups, or looking at the metrological bulletin on TV, and things of that sort. They found a system that provided water films. Not films at the surface of the water boat. Typically, what could have happened is much like an Indian scene where it all happened with an young couple that came to a high priest and said, is it proper that we get married? And of course this had all sorts of economic and political implications. Well, the priest would take a bow and operate this technique. Now, we call this Lecanomancy. Mancy means divination and Lecano means bowl in Greek – Divination from the bowl. He would search for a case which had black films in this surface, a non-colored frame, he would prefer to see something of this sort. Well, there is the married couple and some great Gods, standing close. And there is lot of motion, I will come back to that, and anything could happen when you can have a romantic angle of putting spheres together, where you can have traumatic ends, and so on and depending on what happened the priest presided the wedding which looks like a very random process but this is probably not much worse than the process which we have now-a-days. That would be another talk.

The person who has advanced this field mostly in the last thirty years or so is Cowell Minane. He was born in Poland. He was educated in Switzerland and France and he ended up in California, where he lives now happily. Another remarkable experimentalist of the Plato type. For instance, even a few years ago he was writing regularly in his note book what he has thought. He writes: I have stopped my experiment using these very delicate films. Because there is a signal which perturbs me, I am not sure what it is, but my suspicion is that it is an earthquake. What he was detecting was the San Francisco earthquake which he was detecting from Lahoya at the other end of the state of some 500 miles. Well, we have a lot of admiration for Cowell. He was the man who understood many of these systems. Let me insist on the fact that the things move around. Usually, if you look at the soap bubble just being prepared you will see things moving around. Where does this motion come from? The answer turns out to be very simple. One thing we all know is that if we have a big balloon and we inflate it with hot air it will go up, if the hot air is lighter than the ambient cold air. This is exactly the analog of the balloons. The point is that the black film is very thin, it is very light. Its weight per unit area is very small. While the colored film is 10 times thicker, 10 times heavier, so, indeed we have a situation which is like a balloon. And for that reason this film goes up.

In fact you can do this experiment in another context. Not in warm countries like here but in cold countries if you have central heating or if you have a vertical stove, you will notice that very near the stove valve if you look at it, just tangentially, you see some strange mirage effects. This is exactly that analog. This is an assembly of random balloons being produced by the stove and coming up, in what we call a turbulent mood, very agitated. So we see the analogies between these things and things in very different words.

Turbulence is an interesting feature of these films. Usually the standard way of observing turbulence is to stand on a bridge and look at the river below, choose a moment which is the monsoon season, so that you have lot of water in the river and you notice large eddies from the pyres of the bridge. These eddies flow away and break up in to small eddies, they make all sorts of strange unpredictable things.

That is exactly the same thing with these films. Turbulence is easy in a thick river, essentially because the friction from bottom is not felt. If you observe the river in a season where the bottom is very shallow, and when you have too much friction from the bottom, the eddies die before having had the time to do something interesting. So you want the support of the friction. Now, you see these films are so thin. They are just hanging in thin air. They see very little external friction. So, they are naturally excellent candidates to observe turbulence, turbulences in two dimensional sheets that interested people for twenty years. We did very extensive computer studies about that. But the basic experiment was again a Benjamin Franklin experiment. This time let me be a little bit nationalistic, it was done by Benjamin, some nine years ago. What he does is, he simply moves some sort of pencil to a rectangular film of soap, which he has spread on the large surface, a meter of surface. And this creates a very complicated turbulence behind. Typically the sort of thing which comes from the pencil region is eddies coming one after the other. Sometimes the eddies escape, and go to infinity, you think that film is rolling this way here, rolling this way there. And it's a very complex two-dimensional system which as, I said, has been studied by computer means, before it ultimately got really solved by this simple experiment.

The computer experiments which had been done before at Comain were involving most powerful computer wizards, which was a Cray and the cost of the time which was involved in this was probably in the range of 300,000 dollars. The cost of the experiment of this set up is of the order of 60 dollars. This is a Benjamin Franklin experiment.

I hope that I have given you a little feeling of why there is motion. Now, let me go very briefly to the depth of the film. Usually we see a bubble which hits some corner of the table and bursts. Cowell gave the suspended films which came two little electrodes on the two sides which created a spark in fact simply by using a flash light battery. This little spark would create a hole and this hole Expands —— Expands —— Expands. There's no water left in this hole region and this has been the subject of study for a long time. This is very interesting. It's very analogous to what we have in explosives for instance. If you explode a charge or you explode a nuclear weapon you could have shock waves. This is the analog at a minute scale. Instead of having energies at mega tons of explosives, you have energies which are millions of billions of times smaller. And you can study the same thing. Again Benjamin Franklin.

Well, this brings me to the end of my story. You have seen the birth, mature age and the old age of films. I think this is the conclusion for today. Thank you very much for your attention.

Beyond the Standard Model:
Will it be the Theory of Everything?

Yuval Ne'eman

Tel-Aviv University, Israel

Fig. 1. Dr. B.G. Sidharth, Director, B.M. Birla Science Centre, presenting the Centre's Life Time Achievement in Science Award to Prof. Yuval Ne'eman

Yuval Ne'eman was born in Tel Aviv, Israel (but then Palestine) in 1925. His grandfather, an engineer was one of the sixty six founders of the City of Tel Aviv in 1909. On obtaining the matriculation degree in 1940 in Tel Aviv, he followed his family tradition by studying Mechanical and Electrical Engineering. However in 1940, the young Yuval heard a series of popular lectures on Modern Physics by S. Samboursky of the Hebrew University. This aroused his interest in Physics and he went on to read Eddington's "The Nature of the Physical World". In the meantime he also acquired an encyclopaedic knowledge of history, geography, linguistics and other subjects.

During the study of Engineering, Yuval was also an active member of Haganah for fighting a guerrilla war against the Germans. Immediately after the second world war Yuval was involved in Israel's own long drawn war of independence. His vast military experience got him the job of Director General of Military Intelligence of the Israelian Army in 1955. After 1956 Yuval had

B.G. Sidharth (ed.), *A Century of Ideas*, 87–100.
© Springer Science+Business Media B.V. 2008

a strong desire to return to science and asked for a two year leave of absence from the Army. However at the suggestion of the then Israeli Chief of Staff General Moshe Dayan, he instead took up the Post of the Defence Attache at the Israeli Embassy in London, with the intention of pursuing science at the same time. In the process, Yuval became a part of Abdus Salam's group in 1958. Inspite of heavy work as an Attache, Yuval continued his studies and by October 1960, Yuval achieved his first major result, that of identifying $SU(3)$ in classifying the symmetry of hadrons of strong interaction. Proof for the $SU(3)$ symmetry came in 1964 itself with the discovery of $\Omega-$ Hyperon which was one of the predictions of $SU(3)$. Thus Yuval was one of the pioneers in the progress towards a deeper layer of particles. In 1962 Yuval and Goldberg-Ophir suggested that the baryons themselves were to be made of three even more fundamental particles. Independently similar considerations were worked out by Gell'mann and Zweig.

The 1969 Nobel Prize was given to Gell'mann. Yuval Ne'eman had been bypassed. Ne'eman's contributions in Physics, Astronomy and the philosophy of science continued over the years. Meanwhile in 1961 Yuval submitted his P.hD thesis entitled "Gauges, Groups and an Invariant Theory of Strong Interaction" and returned to Israel to become the Scientific Director of Israel's Atomic Commission. However in September 1963 at the invitation of Murray Gell'mann he spent two years at Caltech, returning in 1965 to become the Head of the Department of Physics at the new Tel Aviv University. Between 1971 and 1975 he was the President of the Tel Aviv University. In 1979, he became the Director of the newly established Sackler Institute of Advanced Studies, a position which he occupied for many years. In between Yuval had several visiting appointments as for example at the University of Texas, Austin and several other national positions in Israel including the top slot at Israel's space organization. He died in 2006.

He was also a Member of several academies and societies for example the Israel National Academy of Sciences and Humanities, the American Academy of Arts and Sciences, the New York Academy of Sciences, the Institute of Physics and the Physical Society (London). His honors and awards include his Weismann Prize for Exact Sciences, the Honorary DSc of the Technion-Israel Institute of Technology, the Albert Einstein Medal and Award, the Honorary DSc of Yeshiva University, New York, the Life Time Achievement of Science Award of the B.M. Birla Science Centre and many more.

Despite his military background, Ne'eman was a very humane person. He explained to me at length, with great objectivity the problem of the Jews. According to him, they are not exceptional people. Rather their persecution over the centuries pushed them towards education and also into trades which other communities were loathe to take up, such as moneylending, professional jewellers and so on. In fact he traced the word jew to jewels and recounted an amusing incident in the USA. His host friend came to know that there was a jew around. So she called him up and asked if he could set right her earring.

Yuval was a great admirer of Mahathma Gandhi, and in his public address wondered how the earth could harbour two such opponents as Gandhi and Hitler at the same time.

Nor was he bitter about being excluded from the Nobel Prize. He told me, "Gell'mann I suppose was given the Nobel Prize because of the much greater background to his work." His encyclopaedic knowledge was evident, whenever he spoke. There are insightful references to various topics which bring to light events and incidents from history and even other topics, including personalities. He shared with Gell'mann, apart from the Eightfold way of particle physics, a propensity for languages. He once recounted an interesting incident. He was walking with Gell'mann somewhere in Israel. Gell'mann turned around and told him, "I hear some people speaking Malayalam (one of the Indian languages)". Clearly this linguistic streak was shared by Gell'mann too.

---

1 Introduction

We learn from Plutarch [1] that "$\pi\lambda\alpha'\tau\omega\nu\epsilon$" $\lambda\epsilon\gamma\epsilon\tau o'\nu\Theta\epsilon o'\nu\alpha'\epsilon\iota\gamma\epsilon\omega\mu\epsilon\tau\rho\epsilon\iota\nu$, that is that "*Plato alleges that God forever geometrizes*". This was a sequel to Democritos's atoms, which were all made of the same *hyle* but in which the four elements corresponded to different perfect geometrical shapes, namely four out of the five perfect polyhedra in three dimensions – (with the fifth polyhedron being assigned to the characterization of the universe). It was also a foreboding of Aristotle's canonization of circular motion. Note, meanwhile, that Aristotle's theories formed a *Theory of Everything*, whose apparent completeness made it possible for it to become part of the Christian (and as a result also of the Jewish and Moslem) canon for a thousand years. One of the instruments through which that gospel was disseminated was Ptolemy's *Almagest* [the name represents the Arabized form of 'the greatest' (collection), that is al, the article 'the' in Arabic, then $\mu\alpha\gamma\iota\sigma\tau\epsilon\,\sigma\iota\nu\tau\alpha\xi\iota\sigma$]. This lecture being presented in India, it is worth recalling an anecdote relating to the *Almagest* which played an important role five hundred and some years ago.

The radius of the earth R was evaluated around 250 BC by Erathostenes of Cyrene, with a precision of better than 0.5%. Knowing that on a given day the sun is at the zenlith (that is at 90° elevation) in Syena, in Southern Egypt (casting no shade inside a deep well), he measured its elevation in Alexandria, that is 90° $-\ \alpha$. Assuming the sun's rays to be parallel (a good assumption, because of its relative distance) you may check and realize that the difference α in elevations is also the difference in latitudes between Alexandria and Syena. By that time, Egypt had been a united kingdom for three thousand years, and the distance δs between the two cities must have been measured quite accurately. The circumference of the earth could thus be evaluated with good precision from $2\pi R = (360/\alpha) \times \delta s$.

About one hundred years later, R was reevaluated by Poseidonius, head of the School of Rhodes and Alexandria, for the star Canopus in the southern skies. For the distance between Rhodes and Alexandria, however, he could not rely on estimates by sailors, who gave him a figure which was too short by some 30%. As a result, Poseidonius's value for R was also too small by 30%. For reasons unknown to me, Ptolemy chose to quote Poseidonius's wrong result in the Almagest, rather than Erathostenes's right one.

Fifteen hundred years later, Christopher Columbus based his proposal on the *Almagest*. With an earth smaller by a third, he evaluated the distance to India, going westward from Portugal, and claimed it to be the shorter route, when compared with Vasco de Gama's trip around Africa. Columbus's proposal was presented as a promising economic venture, namely lower transportation costs for the import of spices from India. Queen Isabella's cabinet submitted the proposal for refereeing to the University of Salamanca. The scholars there were aware of the error in the *Almagest* and had identified its cause, as they now disposed of relatively good maps of the eastern Mediterranean. They thus rightly demolished Columbus's commercial argumentation; moreover, they pointed out that he would never make it to India, since the trip would last six months, instead of Columbus's estimate of three months, and the ships indeed could carry food for just three months. In the film with Gerard Depardieu as Columbus, the picture is distorted: Columbus is credited with the claim that the earth is round, whereas the professors at Salamanca are presented as declaring it to be flat. Moreover, Columbus does not mention the import of spices from India; instead, he keeps mumbling something about a search for 'new worlds to discover'.

What is the lesson? I have pointed out elsewhere [2]–[4] to the *role of research in providing for the introduction of a random element in the evolution of human societies*. Any evolutionary process involves: (1) a mechanism producing random 'mutations' and (2) a process of selection by positive innovation, preserved through some stability criteria. A truly 'important' discovery is one which could not be predicted by extrapolation from the previous stage. Moreover, the evolution of science itself similarly benefits from *serendipity* [5, 6], namely unexpected results popping up in a research program, e.g. Fleming's discovery of antibiotics, after finding the bacteria dead in a Petrie dish whose cover was not well closed. When deciding on a research proposal, it is not sufficient to judge its merits by the expected results as described by the investigator. Other important criteria should include the extent to which the project might explore virgin sectors of phenomena – and the researcher's previous performance, especially in noticing new openings (had he or she been faced with Fleming's dead bacteria, would the only conclusion have consisted in a decision to tighten the lid next time?) A further illustration of the role of the unexpected is that the route to India via the Pacific is indeed utilized nowadays because it is the shortest – for travellers from California, for instance. Thus, Columbus's trip and serendipitous discovery has served – after a few hundred years, to create a market for the product he was advertising in his proposal – trips to India. In any case, although the Salamanca referees were correct, Columbus's connection at the court did manage to ensure that his project be funded – and America discovered.

2 From Leibniz to the Modern Action Principle

We close this anecdotal parenthesis and return to our review of the route
to a "*Theory of Everything*". Our theme of Geometry and the description
of the world now takes us to the seventeenth century and Rene Descartes,
searching for an overall *axiomatic foundation* capable of supporting such a
theory of 'everything' and reformulating geometry – the model for such a log-
ical construction – in a format lending itself to mathematical generalization.
We now come to Leibniz, in the late seventeenth century. It is interesting to
note the nature of the considerations which led him to his invention of calculus – as against Newton's simultaneous creation. For Newton, this involved
the description of velocity and acceleration. Leibniz's opus is entitled "*Nova
Methodus pro Maximis et Minimis*". He regarded physical reality as a strongly
coupled nonlinear system whose stablest solutions are represented by some
maxima or minima. Having no dynamical foundation to build on, he settled
for a maximum – "the best of possible worlds", in the words of Dr Pangloss,
Candide's Leibnizian tutor, in Voltaire's sarcastic treatment (Pangloss per-
sists with this belief even after the worst catastrophes happen to his pupil in
their trip). This is yet another approach to a Theory of Everything, which was
renewed in 1958–1971 in the form of G. Chew's 'Bootstrap'. In this approach
there are no fundamental constituents of matter.

Leibniz's approach also launched the notion of *Invariance* as a basic cri-
terion in building a theory. The idea was developed by the first generation
of Bernoullis (in science) who, influenced by Leibniz, launched the *Calculus
of Variations*. Their Basel colleague Leonhard Euler further improved the
method, which was then formulated as a minimum by Maupertuis, in his
Principle of Least Action. This was no more the best of all possible worlds;
it became the laziest. Lagrange further generalized the principle and freed it
from some irrelevant philosophy which Maupertuis had mixed into it. Finally,
Hamilton and Jacobi developed a different (though closely related) mold. In
our century, Planck's discovery of Quantum Theory further canonized the role
of the Action, now a 'natural' entity and thus a dimensionless quantity, the
basic constituent of physical reality. Schwinger adapted the Action Principle
to the formalism of relativistic quantum field theory and related it to the
S-matrix and finally, Feynman to the path integral.

Although the mold is the Calculus of Variations, the probe is now *Group
Theory*, invented in the first quarter of the nineteenth century by two creative
young men who nevertheless both suffered tragic fates. In Paris, Evariste
Galois, whose papers, written while he was still a high school pupil, were lost
or disregarded by such as Cauchy, Poisson or Fourier, died in a duel at the
age of twenty-one. In Oslo, Henrik Abel died after having eaten straw from
his mattress – out of hunger – one week before the arrival of a letter appoint-
ing him to a chair of mathematics in Berlin. The relevance of group theory
to geometry was pointed out in 1872 by Felix Klein and another Norwegian
algebraist, Sophus Lie, in the *Erlangen program*, launched by Klein in his

professorial inaugural address. The importance of the combined variational application of group theory and geometry to physics was abstracted and clarified by Emmy Noether, a collaborator of both Klein and Hilbert in Goettingen in 1919. In her two theorems, Noether was generalizing from the concrete spectacular applications of symmetry by Einstein, first kinematically (with the Poincare group used globally) in Special Relativity and then dynamically (with local diffeomorphisms for covariance – and after 1928 with the Lorentz group on local frames) in the General Theory. Note that in his 1905 derivation of the Special Theory, Einstein had no inkling of the geometrical nature of what he was doing; it was Minkowski, in his 1908 Cologne address to the German interpretation, with the metric now bearing his name. This was crucial in providing Einsteiln with a foundation that could then be adapted to the treatment of gravity. Classical Mechanics had gone geometrical.

And yet between 1920 and 1970 the physical frontier appeared to have very little to do with geometry – except for the Hilbert space of quantum states, in fact replacing space time as the arena of physics. As to symmetry, between 1925 and 1960 it was generally restricted to the permanently pertinent groups of 3-rotations and Lorentz transformations. Whether they liked it or not, physicists would have to learn to live at least with these, as Hermann Weyl explained [7] in the introduction to the second edition of his *"Group Theory and Quantum Mechanics"*, reacting to the complaints about "the Group Pest".

The adaptation of Quantum Mechanics to the symmetry of Special Relativity in the form of Relativistic Quantum Field Theory (RQFT) scored highly in 1946–48 in Quantum Electrodynamics (QED), but by 1955, the attempts to extend the treatment to Strong and Weak Interactions appeared to have failed. The result – perhaps a somewhat desperate and apocalyptic measure – was Chew's proclamation of the Bootstrap, a theory very much in the spirit of Leibniz's dream. Gone was the challenge of constructing a structural theory; instead, we were told to be satisfied with on-mass-shell physics, reading it out from an experiment, this being the result of a strongly interacting and non-perturbative system.

3 The Standard Model

Around 1961, however, Group Theory returned massively. Hadronic SU(3) and its chiral and spin extensions SU(3) $\otimes$ SI(3) and SU(6) $\otimes$ SU(6) provided an extensive kinematical description of hadron systematics, effective dynamical theories for both low and high energy regimes and a direct lead to the Quark Model as structural foundation [8]–[10].

Throughout the (both geometrically and algebraically!) dark ages of 1920–61, some mathematicians and physicists had nevertheless pursued the unpopular effort. The founder of this 'underground' was Weyl, in his initial 1918 attempt to include electromagnetism in GR, in the guise of a fiber bundle with scale invariance R^1 as the (local) structure group. This failed, but the replacement of R^1 by the compact U(1) as the (local) symmetry of the electron wave function's quantum phase [11] was a clear success, as indicated by its incorporation as an essential element in QED in 1948. Yang and Mills [12] generalized the method in 1953 as a model, for any non-abelian Lie group. Richard Feynman, who had encountered new difficulties when he had turned to the quantization of GR, after his success in QED, now decided to use the Yang-Mills model as a pilot project in that program. Between 1958 and 1962 he first discovered the off-mass-shell loss of unitarity (which was used by Chew as a further proof of the demise of RQFT) and then resolved the difficulty by the invention of ghost fields. B. DeWitt, Slavnov and Taylor, Faddeev and Popov further perfected the method (later reformulated algebraically by Becchi et al. [13] and by Tyutin and reinterpreted geometrically by J. Thierry-Mieg [14]). In 1971, 't Hooft [15] completed the renormalization program for both the symmetric YM theory itself and also for the spontaneously broken case of the (Weinberg-Salam) Electro-weak [16]–[18] gauge theory of $S'U(2) \times U(1)$. Suddenly, RQFT was seen to have scored again. In 1973, Gross and Wilczek, Politzer [19]–[20] (and 't Hooft himself) had all discovered *asymptotic freedom*, thus enabling Weinberg and Fritzsch and Gell-Mann to suggest Quantum Chromodynamics (QCD) [21]–[22] as a fundamental theory of the Strong Interactions. We have since had the Standard Model (SM) as a dynamical grand synthesis, in its nature an atomistic and fully geometrical fundamental theory. The local gauge QFT is built around a Principal Fiber Bundle with space time M as base manifold and G as structure group (= gauge group), with (I', Y' are respectively, the weak isospin and weak hypercharge) $G = SU(3)_{color} \otimes [SU(2)_{I',L} \times U(1)_{Y'}]_{Elweak}$. The electroweak group, however, is broken by a scalar Higgs field $\Phi(x)$ and its hermitean conjugate, with assignments $I' = 1/2, Y' = 1$ for $\Phi(x)$. For the first time since 1915, the geometrization of physics has now encompassed the entire domain of known interactions, outside of GR. All fundamental physics is now geometrical, though on different footings: gravity through classical GR and the SM quantum mechanically, as a QFT.

Note that the enthroning of geometry was not the result of fashion or conditioning. The electroweak theory is a Yang-Mills gauge theory because, in 1957, when the experimentalists set out to measure the coefficients for the contribution of each of the five (S,P,V,A,T) Dirac spinor bilinears, the answer was simply (0,0,0.5, - 0.5,0), i.e. pure $V - A$, the left-handed conserved chiral current, which also explained the non-renormalization of G_V and later fitted PCAC. The strong interaction is a Yang-Mills gauge theory because the model is the only one which produces *asymptotic freedom* in its UV region and could therefore be conjectured to lead to *quark confinement* in the IR zone. The choice of geometry and Plato's vindication all came from experiment.

4 Flavor, the Generations, GUT and the Higgs Sector

The SM is, however, not a closed system with no 'loose ends'. Let us list them.

(1) *The generations*: The physical arena is an Associate Vector Bundle with 15 fermion fields (quarks and leptons) as the fiber, repeating itself thrice. This is an extension of the muon problem, an open issue since the 1947 resolution of the original π, μ confusion. The SM provides no explanation for the existence of the generations.

(2) More specifically, the quark fields are assumed to obey an exact $F = SU(6)_{flavor}$ symmetry. At the level of our hadronic $SU(3)$, this is the solution I suggested in 1964 in my 'Fifth Interaction' paper, [23] to the paradox represented by a Strong Interaction symmetry, nevertheless yielding excellent perturbative predictions. The assumption is that $[F, G] = 0$, with the breaking of the flavor symmetry entering only through the different bare (UV) masses – and being due to other interactions, of a perturbative nature. The combined action of $F \times G$ leads to a matrix of inter-generation Cabibbo-Kobayashi-Maskawa (CKM) mixing, [24, 25] with one complex phase (the CP violation). Note, however, that we also do not have a model for the origin of the bare (UV) masses.

(3) And yet even this is probably not the entire story, as it does not explain the observation that the mass of the top quark is of the order of the W mass, whereas all the five other quark masses may be approximated by zero on that scale.

(4) We have several indications pointing to the existence of a Gauge Unified Theory (GUT) with a simple Lie group U and with G emerging as the subgroup with a residual local gauge symmetry, after the spontaneous breakdown of U. This is indicated (a) by the renormalization group evaluation showing that the three different couplings of the $SU(3)_{color}, SU(2)_{PL}, U(1)'_Y$ all overlap for energies of the order of 10^{16} GeV – and (b) by the fact that within each generation, the summed electric charge of the quarks precisely cancels that of the leptons, an essential condition for the vanishing of renormalization anomalies.

(5) At the same time, we do not know how the breaking of U might be accomplished without causing the hierarchy paradox, i.e. the instability of a sequential set of symmetry breakdowns, due essentially to the QFT problematics of scalar fields in QFT. Candidate answers are supersymmetry or technicolor, with the first as the presently favored solution.

(6) There remains the question of a unification with gravity. Since the advent of supersymmetry, the barrier between space time and internal symmetries has been removed [26]; moreover, for local symmetries, the Kaluza-Klein mechanism has been available since 1920. Supergravity (SUGRA), and especially its $N = 8$ maximal version [27], constructed as a reduction from $N = 1$ in $D = 11$ dimensioins, [28]–[29] seemed to provide the most plausible answer, with its exceptional improvements in renormalizability. Although a theory with the same algebraic advantages, $N = 4$ super-Yang-Mills (SYM)

was proved to have zero radiative corrections (i.e. it is not just renormalizable, it is finite!); $N = 8$ SUGRA is only known to be finite to 8 loops. Related $D = 10$ versions have also been shown to be anomaly-free, for specific gauge groups of rank $\gamma = 16$; this led to the 1984 excitement over strings. This issue is anyhow strongly tied to the question of the existence of Quantum Gravity, which we now discuss.

5 Four Out of Five Ways to Quantum Gravity

I have reviewed elsewhere [30] and rather exhaustively the status of the five possible answers that have been proposed to the riddle of Quantum Gravity. My list comprises,

(1) Leon Rosenfeld's suggestion that Gravity be treated classically, providing the classical background necessary for the masurement apparatus in the Copenhagen interpretation of QM. This has been shown to imply the possibility of violating the Uncertainty Principle through the use of gravity to measure position and velocity.

(2) Roger Penrose's conjecture relating the issue of QG to the collapse of the state-vector, the main residual nonintuitive feature in QM. The latter was blamed by Eugene Wigner and a few others on the incompleteness of QM which does not contain a representation of the human mind as part of the observational setup. In his books, Penrose presents a purported (Godel-type) proof that the human mind is nonalgorithmic and conjectures that the solution to QG can only come after this feature is inserted in the treatment. Although Penrose's books are very well written, I have not been convinced of the juxtapositioin of the two issues, neither have I ever bought Wigner's argumentation in QM. In the latter subject, I do not believe that a quantum measurement implies the presence of a mind – any interaction with a macroscopic body will represent the appropriate irreversible act.

(3) Perturbative QFT fails for Einstein's Lagrangian because of the dimensionality of Newton's constant (as against the dimensionless coupling of a Yang-Mills theory or QED). Renormalizatioin would thus require a new type of counter-term for every order of PT. Adding YM-like terms, i.e. quadratic in the curvature, makes the theory renormalizable, as shown by Stelle [31] and recently reinterpreted by Tomboulis [32]. These authors have, however, also shown that the theory is not unitary. This is due to the emergence of p^{-4} propagators, which contain an on-mass-shell ghost (since they can be written as the difference between two ordinary poles, one of which thus has the wrong sign for its residue). These propagators exist in the theory because it is Riemannian, with the connection being given by the Christoffel formula $\Gamma(x) \sim \partial g$, which results from $Dg = 0$, the Riemannian condition (g is the metric). I have launched a program [33]–[35] in which the high-energy theory is non-Riemannian, i.e. an affine theory such as the SKY [36]–[38] model, but with the double-covering metalinear group $\overline{SL}(4, R)$ as the invariance

group on the local frames. In this theory, a spontaneous symmetry breakdown $SL(4, R) \rightarrow SO(3, 1)$ makes the effective low-energy theory be Riemannian, i.e. Einstein's. The theory is renormalizable [34, 35] but we do not yet have a verdict about its unitarity, due to the use of p^{-4} t4erms in the gauge-fixing part of the quantum Lagrangian.

(4) *Canonical quantization.* Initiated by Dirac in the fifties, this program really took off in 1986, when Abhay Ashtekar [39] rewrote Einstein's theory as a complexified Yang-Mills-like theory, modelled on the apparent complexification of QCD, when the topological (instanton) term $F\Lambda F$ is added to the physical $F\Lambda^*F$, with an imaginary coefficient, as variationally imposed (yielding self-dual or anti-self-dual solutions). Quantization has been achieved in the loop representation [40]. There are some remaining difficulties related to the reality conditions needed to retrieve the physical solutions – and issues of interpretation. The latter are raised by our inexperience with non-perturbative quantization – there is no Fock space and we are handling something which is not 'a set of gravitons'. The loop structure also yields a quantization of area, [41] somewhat in the spirit of Planck's original guess, which was based purely on dimensional arguments. With the advent of GR and QM, however, we recover the Planck length (upto a factor of $\sqrt{2}$) as that length at which the Schwarzschild radius coincides with the Compton wavelength, thus forcing space time to fold up when probed, $(2GM_P/c^2) = (\hbar/M_P c)$, yielding $M_P = (c\hbar/2G)^{1/2}, L_P = (2G\hbar/c^3)^{1/2}$. I have recently shown [42] how these quantities are reproduced in the modern decoherence program, in which the classical background required by the Copenhagen interpretation arises from environmental interactions. Note that the Ashtekar program quantizes GR and does not require it to be unified with other interactions, contradicting the solution we treat next.

6 The Fifth Way: The String and M-Theory

We now come to the fifth solution, which we treat separately because it is also a TOE.

(5) *The quantum superstring.* This is a successful perturbative solution, with the Feynman path integral replaced by a summation over all surfaces. Historically, the hadronic superstring was born when the Leibnizian Bootstrap program inadvertently got itself impregnated with some algebraic semen. Under the taboo laid on QFT around 1958, Gell-Mann invented in 1962 the Matrix Mechanics methodology of Current Algebra for Chiral SU(3). This was further developed by Fubini and his group ('superconvergence') and applied to the Bootstrap by Horn and Schmid in their Finite Energy Sum Rules (with similar results being reached by other groups in Japan and the USSR). Harari and Rosner showed in 1969 that this could also accomodate the quark model, and Harari and Freund added a useful separation between the diffractive and non-diffractive parts of the amplitude. Veneziano's 1967 solution to

the FESR (Dual Models) was shown by Susskind, Nambu and Nielsen and Olesen to represent a string. The removal of a tachyonic parasite state was achieved in 1971 by supersymmetry, first on the string world sheet (Schwarz, Ramond) and later on the embedding 'target' space (Green-Schwarz). As a hadronic theory, it was still plagued by a spin $J = 2$ massless state, until 1974, when first Yoneya and then more directly Schwarz and Scherk, suggested reidentifying the model as a theory of QG. Nevertheless, the program was only really launched some ten years later, after its adoption by E. Witten under the impact of the discovery of the vanishing of chiral anomalies for two $r = 16$ groups [43, 44].

The problem with the quantum superstring (QSS) as a theory of QG is that it is a Theory of Everything, formulated in $D = 10$ dimensions. The quasi-uniqueness of the theory (there are 5 allowed models at the $D = 10$ level), especially as relating to symmetry (either $E(8) \otimes E(8)$ or $SO(32)$) is lost in the myriads of possible Kaluza-Klein compactifications, in the reduction to Minkowski space time. A knowledge of the entire structure is needed because gravity shares the model with all other interactions. In my personal view, the QSS has one strength and one weakness as a candidate meta-theory beyond the SM. Its good point is that like the transition from SR to Galilean physics ($c \to \infty$), from quantum to classical mechanics ($h \to 0$) and from GR to SR ($g_{\mu\nu} \to (1, -1, -1, -1)$), there is a limit ($\alpha' \to 0$) at which we cross from QSS to QFT. The unpleasant feature is that GR is jumped over. The QSS exists in a flat target space; as a matter of fact, this is where it gets its properties (the vanishing of the dilational quantum anomaly). GR is recovered by allowing the target space to curve and imposing the preservation of these properties. Moreover, the graviton continues to exist as a state in the flat limit. Gravity thus takes on the features of a perturbation, a description which, in my eyes, is unbefitting a theory of gravity. Note, however, that there are no wrong predictions that I know of, as resulting from this aesthetic failing. Just an uneasy feeling.

The QSS displays a duality transformation $R \Longleftrightarrow 1/R$, which imply a modification of the Uncertainty Relations for space time [45], $[x, p] = \imath\hbar(1 + \alpha'p^2 + \cdots)$. Since 1995, the TOE program has embarked on a new path, spanned by dualities. First, it was shown that the heterotic string (one of the five allowed bases) has a solitonic quantum solution which is a 5-dimensional extendon (or p-brane in the more popular nomenclature, which I was unable to overturn); moreover, taking the 5-extendon as fundamental, one finds it has the heterotic string as solution! [46]. This seems to represent a reminder of the string's origins, the Horn-Schmid Bootstrap. It is a duality transformation which is also somewhat akin to the electric/magnetic transition. The above $R \leftrightarrow 1/R$ also implies a 'strong $\leftrightarrow$ weak' transition in the coupling, a very useful feature in a strong coupling theory situation.

The other new departure is an extension of the string by one more spatial dimension (in both target space and the world surface) leading to a quantum membrane [47], with a QFT limit reproducing 11-dimensional supergravity.

This 11-dimensional 'M-theory' also manages to relate to all five QSS theories, through a chain of duality transformations! I can thus point to four promising features: (a) further unification, (b) calculability (through the strong/weak transition), (c) a return of $N = 8$ supergravity, with its precise components, as a unifying QFT, and (d) the possibility that we might be close to the elaboration of a new fundamental generating principle, realized by the theory – in the way that covariance and equivalence generate GR. The lack of such a principle has been another suspicious feature in the QSS, which has up to the present appeared to emerge from a whim or from an arbitrary choice. The new principle might be a variation on the bootstrap theme, in terms of d-extendons (rather than hadrons) generating each other, for instance. This would be a fully geometrical theory – which is a good omen, considering past experience. The QSS has had a history of 'highs' and 'lows', and this is the latest 'high'. Will it last?

7 Non-commutative Geometry

I would like to add a few words on the new mathematical discipline of non-commutative geometry [48], which appears to apply to our discussion. It is being used by Sidharth and others. It represents a method of assigning to discrete spaces notions (such as a metric, i.e. distances) developed in continuous manifolds. It is strongly inspired by physics, the generalization being based on the definition of a space through the functions and Hilbert spaces it can carry. It appears to me at this point to be applicable in three main directions:

(1) Replacing space time by 'fuzzy' spaces, whose volume is finite. This could yield a new mode of renormalization [49]. It has worked in some examples; however, it seems as unjustifiable as an ordinary cutoff, which achieves the same result.

(2) Providing a geometric derivation for the spontaneous symmetry breakdown of a YM gauge theory. It yields a superconnection [50], a supermatrix whose even submatrices (along the diagonal) are valued over the (Grassmann-odd) gauge field one-forms, and whose odd sector is valued over the Higgs field zero-forms. The overall grading is thus odd everywhere. The carrier space can be graced by any relevant feature (e.g. chirality) and does not have to be tied to quantum statistics, as in ordinary supersymmetry. Connes and Lott [51] first showed that they could reproduce the electro-weak theory in this manner (by starting with a $Z(2) \otimes E^4$ base space. E^4 is euclidean flat space time, $Z(2)$ is a two-point space, these being L and R, the two chiralities). It was then shown [52] that with slightly more sophisticated geometrical assumptions, one reproduces the superconnection version [53] of my 1979 model [54,55], an irreducible superunification of the electro-weak theory. I have recently suggested a model of this type for gravity [56] (in the perturbative program which I described in the section Four out of Five ways to Quantum Gravity). Connes has extended his original treatment to a unification of the electro-weak with gravity [57].

(3) Geometrization of quantum mechanics: My student Atzmon [58] has studied the geometrization of a series of potentials, oscillators, etc. – with their energy levels. Very roughly, the NCG 'distances' are inversely proportional to the transition probabilities. My hope is that it may be possible to explain the non-intuitive features of QM with these methods: suppose, for instance, that we could show that the NCG 'distance' between the two particles in EPR is null – wouldn't this be marvellous for our peace of mind?

References

1. Plutarch, *Quaest.* Plat.
2. Y. Ne'eman, *Acta Sci. Venezolana* **31**, 1–3 (1980).
3. Y. Ne'eman, *Metab. Pediat and Syst. Ophthalmology*, **11**, 12 (1988).
4. Y. Ne'eman, in *Soft Order in Physical Systems*, (Proc. Les Houches, 1993), (eds.) R. Bruinsma Y. Rabin, Plenum Press, New York (1994), pp. 223–228.
5. Y. Ne'eman, *Proc. Kon. Ned. Akad.* v. Wetensch. **96**, 433 (1993).
6. A. Kantorovich and Y. Ne'eman, *Stud. Hist. Phil. Sci.* **20**, 505 (1989).
7. H. Weyl, *The Theory of Groups and Quantum Mechanics*, 2nd (revised) German edition (1930), English translation (Dover, Inc, London).
8. Y. Ne'eman, *Nucl. Phys.* **26**, 222 (1961).
9. M. Gell-Mann and Y. Ne'eman, *The Eightfold Way*, W.A. Benjamin Publishing Co., Reading, MA (1964).
10. Y. Ne'eman, *Algebraic Theory of Particle Physics*, W.A. Benjamin Publishing Co., Reading, MA (1967).
11. H. Weyl., *Z Phys.* **56**, 330 (1929).
12. C.N. Yang and R.L. Mills, *Phys. Rev.* **95**, 631 (1954); **96**, 191 (1954).
13. C. Becchi, A. Rouet and R. Stora, *Ann. Phys.* **98**, 287 (1976).
14. J. Thierry-Mieg, *J. Math. Phys.* **21**, 2834 (1980).
15. G. 't Hooft, *Nucl. Phys.* B, **35**, 167 (1971).
16. S. Glashow, *Nucl. Phys.* **22**, 579 (1961).
17. S. Weinberg, *Phys. Rev. Lett.* **19**, 1264 (1967).
18. A. Salam, in *Elementary Particle Theory*, Svartholm (ed.), N., almquist Publications, Stockholm (1968).
19. D.J. Gross and F. Wilczek, *Phys. Rev. Lett.* **30**, 1323 (1973).
20. H.D. Politzer, *Phys. Rev. Lett.* **30**, 1346 (1973).
21. S. Weinberg, *Phys. Rev. Lett.* **31**, 494 (1973).
22. H. Fritzsch and G. Gell-Mann., in *Proc. XVIth I.C.H.E.P.* Chicago, IL, **2**, 135 (1972).
23. Y. Ne'eman, *Phys. Rev.* B **134**, 1355 (1964).
24. N. Cabibo, *Phys. Rev. Lett.* **10**, 531 (1963).
25. M. Kobayashi and K. Maskawa, *Prog. Theo. Phys*, **49**, 282 (1972).
26. R. Haag, J.T. Lopuszanski and M. Sohnius, *Nucl. Phys.* B **88**, 257 (1975).
27. Y. Ne'eman, *Aspen Inst. lecture*, June (1976), unpublished.
28. E. Cremmer and B. Julia, *Nucl. Phys.* B **159**, 141 (1979).
29. E. Cremmer and B. Julia and J. Scherk, *Phys. Lett.* B **76**, 409 (1978).
30. Y. Ne'eman, in *Proc. Second Mexican School of Grav. and Math. Phys.* Tlaxcala (1996), to be published.

31. D. Stelle, *Phys. Rev.* D, **16**, 953 (1977).

32. E.T. Tomboulis, *Phys. Lett.* B **389**, 225 (1996).

33. Y. Ne'eman and Dj Sijacki, *Phys. Lett.* B **200**, 489 (1988).

34. C.Y. Lee and Y. Ne'eman, *Phys. Lett.* B **242**, 59 (1990).

35. C.Y. Lee., *Class. Quantum Grav.* **9**, 2001 (1992).

36. G. Stephenson, *Nuo. Cim.* **9**, 263 (1958).

37. C.W. Kilmister and D.J. Newman, *Proc. Cam. Phil. Soc.* **57**, 851 (1961).

38. C.N. Yang, *Phys. Rev. Lett.* **33**, 445 (1974).

39. A. Ashtekar, *Phys. Rev. Lett.* **57**, 3344 (1986).

40. C. Rovelli and L. Smolin, *Phys. Rev. Lett.* **61**, 1155 (1988).

41. C. Rovelli and L. Smolin, *Nucl. Phys.* B **442**, 593 (1995).

42. Y. Ne'eman, *Phys. Lett.* A **186**, 5 (1994).

43. M.B. Green, J.H. Schwarz and E. Witten, *Superstring Theory*, Cambridge University Press, Cambridge, U.K. (1987).

44. J.H. Schwarz (ed)., *Superstrings: The First Fifteen Years of Superstring Theory*, World Scientific Publications, Singapore (1958) 2 Vols.

45. E. Witten, *Physics Today*, April 24–30 (1996).

46. M. Duff, *Class. Quantum Grav.* **5**, 189 (1988); A. Strominger, *Nucl. Phys.*, B **343**, 167 (1990).

47. E. Bergshoeff, E. Sezgin and P.K. Townsend, *Ann. Phys.* **185**, 330 (1988); P.K. Townsend, *Phys. Lett.* B **350**, 184 (1995).

48. A. Connes, *Publ. Math* IHES, **62**, 257 (1985).

49. J. Madore, *Class. Quantum Grav.* **9**, 69 (1992).

50. D. Quillen, *Topology*, **24**, 89 (1985).

51. A. Connes and J. Lott., *Nucl. Phys.* (Proc. Suppl.) B **18**, 29 (1990).

52. R. Coquereaux, R. Haussling, N.A. Papadopoulos and F. Scheck, *Int. J. Mod. Phys.* A **7**, 2809 (1992).

53. S. Sternberg and Y. Ne'eman, *Proc. Nat. Acad. Sci. USA*, **87**, 7875 (1990).

54. Y. Ne'eman, *Phys. Lett.* B **81**, 190 (1979).

55. D.B. Fairlie, *Phys. Lett.* B **82**, 97 (1979).

56. Y. Ne'eman, to be published in *Proc. 7th Marcel Grossmann Symp.* Jerusalem (1997).

57. A. Connes, *Gravity coupled to matter etc.*, hep-th/9603053.

58. E. Atzmon, *The associated metric for a particle in a quantum energy level*, to appear in *Found. Phy.* (1998).

Living Joyfully with Complexity in Chemistry and Culture

Roald Hoffmann

Cornell University, Ithaca, NY, U.S.A

Fig. 1. Roald Hoffmann delivering the B.M. Birla Memorial Lecture

Roald Hoffmann was born in Zloczow, Poland in 1937. Early in life, he had to face persecution from the Nazis. He however managed to escape with his family and arrived in the United States in 1949. He graduated from Stuyvesant High School, Columbia University and obtained his PhD in 1962 from Harvard University, working with W.N. Lipscomb and Martin Gouterman. After a brief stint at Harvard as a Junior Fellow from 1962 to 1965, he joined Cornell University where he has been the Frank H.T. Rhodes Professor of Humane Letters and Professor of Chemistry.

Professor Hoffmann is a member of the United States National Academy of Sciences, The American Academy of Arts and Sciences, and the American Philosophical Society. He has been elected a Foreign Member of the Royal Society, the Indian National Science Academy, the Royal Swedish Academy of Sciences, the Finnish Society of Sciences and Letters, the Russian Academy of Sciences and the Nordrhein Westfallische Academy of Sciences. He has received numerous honors, including the Life Time Achievement in Science

B.G. Sidharth (ed.), *A Century of Ideas,* 101–108.

Award of the B.M. Birla Science Centre and over twenty five honorary degrees. He is the only person ever to have received the American Chemical Society's awards in three different specific subfields of Chemistry – the A.C. Cope Award in Organic Chemistry, the Award in Inorganic Chemistry, and the Pimentel Award in Chemical Education, as well as two other ACS awards. In 1981, he shared the Nobel Prize in Chemistry with Kenichi Fukui.

In more than four hundred and fifty articles and two books Professor Hoffmann has thrown a new perspective to look at the geometry and the reactivity of molecules from Organic to Inorganic to infinitely extended structures.

In recent times Professor Hoffmann has looked at the electronic structure of extended systems in one, two, and three dimensions. Frontier orbital arguments find an analogue in this work in densities of states and their partitioning. An especially useful tool, the COOP curve, has been introduced by the Hoffmann group. This is the solid state analogue of an overlap population, showing the way the bond strength depends on electron count. The group has studied molecules as diverse as the platinocyanides, Chevrel phases, transition metal carbides, displacive transitions in NiAs, MnP and NiP, new metallic forms of carbon, the making and breaking of bonds in the solid state and many other systems. One focus of the solid state work has been on surfaces, especially on the interaction of CH4, acetylene and CO with specific metal faces. The group has been able to carry through unique comparisons of inorganic and surface reactions.

He is a multi-dimensional character. Not just a research scientist, he has also been involved in pedagogy and popularization. He participated in the production of a television course on Chemistry. This twenty six episode series was developed at the University of Maryland. Professor Hoffmann was the presenter and narrator of the series. He has also written popular articles as well as thought provoking articles on Science and even the Arts including poetry. In fact in 1993 the Smithsonian Institution Press published "Chemistry Imagined" which was a collaboration with artist Vivian Torrence on Art, Science and Literature. A play, "Oxygen" written with Carl Djerassi had its premier at the San Diego Repertory Theatre in 2001 and had productions at the Riverside Studios in London and Wurzburg and Munich for the German version, in the fall of the same year. This play has been broadcast by BBC World Service and West German Radio and has been published in English and German translations. These are but a few examples of Professor Hoffman's versatility.

He is a very humane person. This trait has undoubtedly been enhanced by his traumatic experience in the Nazi period. In conversation he described at length his escapade, using false names and passports. It is nothing short of a real life thriller. His great love for students and the amount of time and patience he can devote to them is also very touching, as was demonstrated at Hyderabad. He not only patiently gave his autographs to the huge crowd of students, but also illustrated each autograph with one of his typical chemical diagrams.

Though soft spoken, Professor Hoffmann is also outspoken. He said that if he were to meet the Prime Minister (of India), he would tell him that rather than all the esoteric research which is going on in the laboratories, India should tie up with a country like Columbia to develop a much needed malaria vaccine. He added that such a vaccine could be developed in about six months in the United States, but it won't happen for all the wrong reasons. Malaria is a third world disease and there isn't enough money to be made out of such a vaccine.

I am pleased to be in the B.M. Birla Science Centre, at the invitation of Dr. B. G. Sidharth. The Birla Science Centre is dedicated to the widest possible dissemination of knowledge. I am very pleased to be here because I myself believe that we as scientists must be dedicated in the widest possible sense to talking about science to the general public, not only to our students and future colleagues and competitors in science.

There are many reasons for speaking about science to the general public. One motivation could be to attract more people to our profession. More important is that it is impossible for a democratic society such as India to function without the broadest possible awareness of some of the basic ideas of science by the general public. Scientists form only about one percent of our population. Research is possible only when the other 99% of society understand what scientists do. People ultimately make the decisions. They may seek the advice of experts, but experts can be martialled on the side of any issue in the world. It is important that people themselves learn at least some basic ideas of science, so that they can judge the words of experts and listen to them critically, as well as to the words of politicians. It is very important for the functioning of a democratic society that people know about science.

There is another reason, a psychological one. If we do not know how the world around us works, we create, in the tradition of human beings over ages, mysterious explanations and superstitions around the workings of that world. In the old days those things were created around the motion of the planets, around eclipses, comets and other phenomena – what is interesting today is that science and technology have surrounded us with all kinds of things that we don't understand. Do you know what goes on inside a CD player or inside your modern automatic camera? Those mysteries can well separate us from the things that we use, and so soon we are alienated, in the psychological sense, from the world around us.

It is in the spirit of this that I would like to tell you, speaking very much to the young people in the audience and to people who are not at all in my profession, something about chemistry. But what I will tell you about chemistry is not what you will see in a normal textbook. It is a kind of reflection on chemistry or an examination of several cross sections of chemistry.

Chemistry may not sound interesting, because it is in the middle. We don't have the infinitely big and we don't have the infinitely small, we only have a piece of life. But the word interest, if you look at its etymology, comes from the Latin (and then eventually probably from the Sanskrit) inter-esse; "inter" is between and "esse" is to be in Latin. So to be in between is to be potentially interesting. The cosmology of galaxies and the nature of elementary particles will not create a new pigment in a dye. You worry about whether a certain molecule can affect you or not because the molecule is on the scale of molecules in us. Chemistry is interesting. It is in the middle, it is on the human scale, it concerns people.

The first description of chemistry, the first of at least three that I would give you, is one that could have been given five hundred years ago. Chemistry is the art, craft and business of substances and their transformations. This definition predates science. I assure you there was chemistry, not only in our bodies, but chemistry done by human beings, before there was science. Let me give you two examples just from the culture around you – Tapioca/manioc is a substance that has to be processed by boiling in order to remove a poisonous substance in it before it can be eaten. That is chemistry. One of the most beautiful and useful dyes in world culture is indigo. Before people even learned how to make indigo in the laboratory that material was processed from a plant of the pea family and made into a beautiful dye in most tropical cultures. People processed manioc and indigo without waiting for chemists and laboratories, learning from many years of experimentation.

In order to show you the essential transformation at the heart of chemistry, what I would need to do is an experiment, because that is the heart of chemistry. It is some sort of change in some substance, but I did not know if I could do an experiment here. For example let us take a bromine and aluminum. If we put aluminium into bromine absolutely everything that you expect of chemistry takes place – foul odors, smoke, fire, if not an explosion. This is change before your eyes, from the reddish brown liquid bromine and the beautiful silvery metal of aluminium we get an aluminium bromide which is a white powder. This shows what chemistry is about.

The changes that take place are obviously not always so violent or quick as this one. If a neighbor next to you is awake (or alive, even if he is not awake), he is a wonderful example of chemistry at work. Proceeding much more slowly and much more quickly than the reaction we've seen, the enzymes in our bodies are transforming other molecules at the rates of millions of molecules per second. There are vast changes going on in us. Our kidneys, for instance, are processing a pound of bicarbonate every day. Incredible chemistries are taking place within us.

One interesting consequence of chemistry being about substances and their transformations is in a perception of chemistry that follows. In a comic book, Donald Duck comes in and says, "Hi what's cooking?" and his nephews say, "We don't know. We are playing with our chemical stuff." And then he says, "Why don't you pour this stuff in?" There follows the obligatory explosion, the necessary bump on his head. More interesting is the next panel, in which Donald suggest that the nephews mix CH_2 with NH_4. Peter Gaspar and George Hammond brought this strip to my attention; in a paper of theirs on CH_2-like molecules they simply said, "Some experiments on CH_2 suggested in the literature have not yet been tried." The reference was to Donald Duck's Walt Disney comics! Perhaps one of you can do it, it's not an easy experiment.

Consider next the Greek alchemical manuscript illustrating the principles of alchemy. Most scientists, especially chemists, have a rather ambiguous picture of alchemy. Alchemy was a philosophy associated with chemical experimentation that arose in a number of cultures – in China, in India, in Egypt, Greece and Europe in medieval times. In the manuscripts there is a picture of a swan that is biting its own breast and some oils in a chemical flask. In another Alchemical illustration, there is the wedding of a king and a queen by a bishop, but on the side, unlike any wedding that you have seen or are likely to see, people are doing chemical experiments. There is obviously something symbolic being communicated.

Many things came from alchemy: in Europe the making of the strong mineral acids, of sulphuric acid, nitric acid and hydrochloric acid; much of the shape of the glass vessels in alchemical illustrations is not that different from things that we see in a laboratory today. The philosophy of alchemy is change. What kind of change? The change of a sick person to a healthy person. The change of a base metal like lead, to a noble metal like gold. Perhaps a psychological change in a person performing the experiments themselves.

Modern scientists would like to take what the alchemists gave us. They would forget about the underlying philosophy. And laugh a little nervously at the kind of the dishonesty that inevitably accompanied something like making gold out of lead. I would say you can't do that – they are all tied together. What I think is interesting here is that the philosophy of change came first, when it wanted to get into people's souls and hearts, and looked around in the world for something which really represented a change of people – it found Chemistry. Chemistry was being used by a philosophy as a metaphor for change. This is very interesting.

There is a painting by a Dutch painter of 1570, Jan van der Straat. It is now in Florence, and it represents a late Alchemical Laboratory in the court of the Duke of Florence. It shows the patron Duke doing experiments in his newly commissioned laboratory. The woman in the center, holding a flask, is Bianca Capello, the second wife of the Duke. Behind him is a figure that would be recognizable to anybody here, and that is the Master Alchemist. He is the Director of the Laboratory. He is doing nothing, telling others what to do. Around them are figures doing all the work – the graduate students, as recognizable today as then.

This is a wonderful illustration of the eternal sociology of science. But there is a difference between 1570 and our times. When the Duke of Lawrence wanted to have himself, his wife, and his courtiers to be painted in an official portrait, a fun portrait to be sure but still an official one, he dressed up in the clothes of a chemist and did some chemical experiments. Can you imagine the President of the United States or the Prime Minister of India doing that today?

Now something has happened in the last two hundred years – we have learned to look inside the innards of the beast, where the substances are changing. It is in the nature of curious human beings to try to understand what happens when aluminium is placed into bromine and it changes in some way. We have developed the tools for looking inside matter. But something I must tell you right away, there are no microscopes to do this. You can't see molecules, except in some special circumstances. All of the beautiful structure of Chemistry was developed as a kind of knowing without seeing, slowly and laboriously formulated by human beings and their tools, building slowly a body of knowledge of what is inside. We now know that at the microscopic level in substances are atoms, and much more important than atoms – persistent groupings of atoms called molecules.

Chemistry is still the art, craft, business and now science of substances and their transformations. But it is also the same art, craft and science of molecules and their transformations. And any chemist today thinks both micro and macroscopically.

I want to represent some of these molecules for you. So I will show you some of them, some of the simplest possible ones that you can build from carbon or hydrogen atoms. At least two of them are quite familiar to us: Methane – this is the main component of natural gas. You also see propane, which is used in heating in various ways, and in between them, ethane. These are the three simplest hydrocarbons.

I have represented these molecules not in one but in three different ways, which are recognizable to you as a chemical structure, as a ball and stick model, and as something a chemist would call a space – fitting model. Why do I show them in three ways? This has something to do with the communication between scientists and people outside of science, especially people in the Arts and Humanities. You see, scientists have given the world the impression that they have a strangle hold on reality, that they really know what's in there and in the world. That is why TV ads or placards show men in white coats who are telling us what someone would like us to believe is true.

Now the reality is ... that there is an underlying reality, there is a methane molecule, there is a propane molecule. But when I draw it, I am representing reality and communicating to someone else the nature of that molecule. I am very much engaged in representing things and ideas. And there isn't only one way to do that. Reality is objective, representation of reality is subjective – I choose the representation appropriate to the act of communicating my goals and intent, and to the receiver of my message. Sometimes I want to show the chemical structure, that is sufficient. And sometimes I am interested in the shape of the molecule. We do this in chemistry without thinking but it is important to realize how subjective our scientific representations are, because...people in the humanities and the arts are always representing things in many different ways. Is there one way to write a poem about the end of love? You can write a thousand poems and the next one will not be superfluous. Admitting that there are different representations in science is not at all a weakness – it is something of material and spiritual value, that builds a bridge between us and the people in the humanities.

There are more complicated molecules, such as thalidomide. In the 1960s a German chemical company put this molecule out in the market as a sleeping pill and in fact directed its marketing to pregnant women. In fact, the compound was teratogenic, the source of about 10,000 malformed births, mostly in Europe. The interesting thing is that the same molecule, which without doubt has caused incredible pain and suffering, has also a beneficial side. It is a proven therapeutic agent against a form of leprosy. It looks like a promising agent for the treatment of HIV infections and a number of other syndromes. One and the same molecule is both good and bad for people.

The human mind has a lot of difficulty with this idea, reflecting our own ambiguities on good and evil. Which leads me to the question: Are there good and evil molecules? No, there are no good and evil molecules, only good and evil people. Does the society have the right to restrict the production of a molecule even though the molecule is not good or evil by itself? You can argue with me, but I would say that indeed a society has the absolute right to restrict the production of a molecule.

Very often scientists avoid ethical discussions. In a caricature, they might say, "I am just making this molecule; it's not my responsibility to worry about what use you or someone else puts it to." If not I, who then? The world is made of such excuses, and less ethical people are waiting to use such scientists who refuse to think about the ethical consequences of what they do. I think it is a social responsibility of scientists to worry about the consequences of their actions, the molecules they make, and the techniques that they use, even if it is a danger to their jobs and to their own well-being.

Let's talk about the beauty in molecules – even simple-looking molecules, in the shape of Platonic solids – tetrahedral, cubes and even football shapes. These molecules are simply beautiful, beautifully simple, but devilishly hard to make. Except the last one. There is a remarkable irony in this, which serves as an inspiration to everyone working in science. Of all these molecules, by far the easiest to make is the one that was made last of all, and that's the football shaped one. What other things are waiting there to be made?

These molecules project their beauty, shine like a laser beam into our soul. When we see them we are happy. We are happier if we can make them. But wait a moment, simplicity is not all there is to beauty. There are more complicated molecules – take the oxygen carrier in our blood, hemoglobin. It looks like... worms doing a dance. Whatever this molecule is, with its 9,500 or so atoms, is in terms of a scale of simplicity and complexity, relative to the tetrahedron or a cube, light years away. This incredible molecule, with its four essential iron atoms, is certainly beautiful, on every account. But it has nothing to do with simplicity.

The beauty of hemoglobin resides in its function, of carrying oxygen to the lungs, and the way it is suited for it. I would like you to reflect on the fact that complexity is necessary to do things. A human body is not as simple as a liter of petrol. The human body runs lots of chemical reactions at the same time – breathing in, carrying the oxygen, getting the oxygen to the muscle cells, carrying the wastes away, there are at least 10,000 chemical reactions going on in us, and going on very quickly. You need complexity in order to do anything of value in this world. We have some trouble with this notion, probably because our mind, by itself a complex structure, somehow has evolved to favor simplicity, a weakness. This is something which politicians know well; their propaganda takes advantage of this weakness of ours.

Chemists can make structures that are simple, and they can also make molecules which are complex. It is an interesting kind of building we do – it's not at all like building a marble structure. Instead, we mix some chemicals and apply a source of energy, heat or light. Then we let go, and, incredibly, 10^{23} molecules colliding randomly inside a flask create what we want. With a little bit of design, and some luck.

Now there are structures that human beings build on a scale 12 orders of magnitude up from molecules. This is monumental architecture. What does it have to do with chemistry? Well, it is also building. It also takes money, takes talent, ergo human beings, all these things. Buildings and molecules are objects of human creation. What is interesting is that the structures that human beings have chosen to build in this world reflect some of the same questions of simplicity and complexity that are there in the world of molecules. The Taj Mahal of Agra, a high point of Mughal Architecture in India is clearly an expression of an aesthetic in which simplicity is valued. That is obvious in the dominant bilateral symmetry of the lovely structure. Though if you look at some details of the stone tiling and grillwork in the structure, you see tension, the juxtaposition of two different patterns. Symmetry sets repose, but interest is created by asymmetry. The towers of Chalukyan temples of the eighth and ninth centuries provide a good example of this.

I have come to the end of my second cross section of chemistry. As you've seen, this art, craft, business, and eventually science is firmly embedded in culture. Chemists have contributed their skills to the masterpieces of world art; art asks some of the same philosophical questions that chemistry does. Scientists take great risk in evading social responsibility for the magnificence of their creation. It is only by seeing both art and science as firmly embedded in our society, in our economy, in our culture, neither shirking ethical considerations, that we can move both forward. Together.

A Confrontation with Infinity

Gerard't Hooft

Institute for Theoretical Physics, University of Utrecht,
The Netherlands

Fig. 1. Gerard't Hooft delivering the B.M. Birla Memorial Lecture

Gerardus 't Hooft was born at Den Helder to a family with a distinguished record in science. In 1953 his grand-uncle, Frits Zernike had earned the Nobel Prize for the invention of the phase contrast microscope, while his uncle Nicolaas Godfried van Kampen was professor of theoretical physics at the University of Utrecht.

Gerardus spent his childhood in the Hague with his parents and family. When his father tried to get him interested in engineering, he did not evince any enthusiasm, saying "I want to investigate nature and discover new things."

He started his affair with the piano at the age of ten, something he has kept up over the years. At age 16 he participated in the Dutch National Math Olympiad, going to the next round in Utrecht. Though he felt that he had not done well, he came amongst the first ten participants, obtaining the second prize. After passing High School in 1964 he went to the State University of Utrecht, where he was near an Institute of Theoretical Physics. This fostered

B.G. Sidharth (ed.), *A Century of Ideas*, 109–121.
© Springer Science+Business Media B.V. 2008

his desire to go into the exciting problems of elementary particles. Here, for his undergraduate thesis he was supervised by Martinus Veltman. He started his PhD work in 1969, again under the supervision of Prof. Veltman. During this time the renormalization of Yang-Mills fields caught his fancy. He encountered and circumvented the various difficulties in the renormalization of these theories, catching world wide attention with his second paper. He obtained his PhD in 1972, the same year that he got married. Thereafter the t' Hoofts went to CERN, Geneva where they were joined by Prof. Veltman also.

In 1974 they returned to Utrecht. Subsequently Prof. t' Hooft was invited at guest positions at Harvard and Stanford. He spent much time and energy trying to unravel the quark confinement problem. By early 1980s the contours of a possible solution had become clear: QCD could be treated numerically using lattice cut offs.

Prof. t' Hooft continued his intensive researches into one of the final remaining problems, namely a Quantum Mechanical treatment of gravitation. In 1999, along with his former supervisor Martinus J. Veltman, t' Hooft got the Nobel Prize in Physics.

Prof. t' Hooft is very well organized, mentally and physically and very soft spoken. His lectures are packed with insights. I have had the pleasure of meeting him a number of times in different countries and it is always a learning experience. He is interested in human affairs too and has kept up his association with the piano. Once he told me, "When I drink three glasses of wine, I can still drive the car. But after one glass of wine I cannot play the piano."

Amongst his many distinctions and achievements was the Life Time Achievement in Science Award of the B.M. Birla Science Centre.

1 Appetizer

Early attempts at constructing realistic models for the weak interaction between elementary particles were off-set by the emergence of infinite, hence meaningless, expressions when one tried to derive the radiative corrections. When models based on gauge theories with Higgs mechanism were discovered to be renormalizable, the bothersome infinities disappeared-they cancelled out. If this success seemed to be due to mathematical sorcery, it may be of interest to explain the physical insights on which it is actually based.

2 Introduction

In this lecture I intend to reflect on the efforts that were needed to tame the gauge theories, the reasons for our successes at this point, and the lessons to be learned. I realize the dangers of that. Often in the past, progress was made precisely because lessons from the past were being ignored. Be that as it may,

I nevertheless think these lessons are of great importance, and if researchers in the future should choose to ignore them, they must know what they are doing.

When I entered the field of elementary particle physics, no precise theory for the weak interactions existed [1]. It was said that any theory one attempted to write down was non-renormalizable. What was meant by that? In practice, what it meant was that when one tried to compute corrections to scattering amplitudes, physically impossible expressions were encountered. The result of the computations appeared to imply that these amplitudes should be infinite. Typically, integrals of the following form were found:

$$\int d^4 k \frac{Pol(k_\mu)}{(k^2 + m^2)[(k + q)^2 + m^2]} = \infty, \tag{1}$$

where $Pol(k_\mu)$ stands for some polynomial in the integration variables $k = \mu$. Physically, this must be nonsense. If, in whatever model calculation, the effects due to some obscure secondary phenomenon appear to be infinitely strong, one knows what this means: the so-called secondary effect is not as innocent as it might have appeared – it must have been represented incorrectly in the model; one has to improve the model by paying special attention to the features that were at first thought to be negligible. The infinities in the weak-interaction theories were due to interactions from virtual particles at extremely high energies. High energy also means high momentum, and in quantum mechanics this means that the waves associated with these particles have very short wavelengths. One had to conclude that the short distance structure of the existing theories was too poorly understood.

Short distance scales and short time intervals entered into theories of physics first when Newton and Leibniz introduced the notion of differentiation. In describing the motion of planets and moons, one had to consider some small time interval Δt and the displacement Δx of the object during this time interval [see Fig. 2(a)]. The crucial observation was that, in the limit $\Delta t \to 0$, the ratio

$$\frac{\Delta x}{\Delta t} = v \tag{2}$$

makes sense, and we call it "velocity." In fact, one may again take the ratio of the velocity change Δv during such a small time interval Δt, and again the ratio

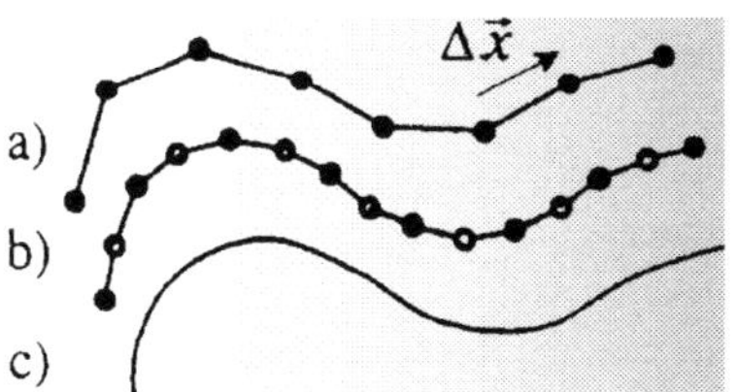

Fig. 2. Differentiation

$$\frac{\Delta \boldsymbol{v}}{\Delta t} = \boldsymbol{a} \tag{3}$$

exists in the limit $\Delta t \to 0$; we call it "acceleration." Their big discovery was that it makes sense to write equations relating accelerations, velocities, and positions, and that in the limit where Δt goes to zero, you get good models describing the motion of celestial bodies [Fig. 2(c)]. The mathematics of differential equations grew out of this, and nowadays it is such a central element in theoretical physics that we often do not realize how important and how nontrivial these observations actually were. In modern theories of physics we send distances and time intervals to zero all the time, also in multidimensional field theories, assuming that the philosophy of differential equations applies. But occasionally it may happen that everything goes wrong. The limits that we thought to be familiar with do not appear to exist. The behavior of our model at the very tiniest time and distance scales then has to be reexamined.

Infinite integrals in particle theory were not new. They had been encountered many times before, and in some theories it was understood how to deal with them [2]. What had to be done was called "renormalization." Imagine a particle such as an electron to be something like a little sphere, of radius R and mass m_{bare}. Now attach an electric charge to this particle, of an amount Q. The electric-field energy would be

$$U = \frac{Q^2}{8\pi R}, \tag{4}$$

and, according to Einstein's special theory of relativity, this would represent an extra amount of mass, U/c^2, where c is the speed of light. Particle plus field would carry a mass equal to

$$m_{phys} = m_{bare} + \frac{Q^2}{8\pi c^2 R}, \tag{5}$$

It is this mass, called "physical mass," that an experimentor would measure if the particle were subject to Newton's law, $\boldsymbol{F} = m_{phys}\boldsymbol{a}$. What is alarming about this effect is that the mass correction diverges to infinity when the radius R of our particle is sent to zero. But we want R to be zero, because if R were finite it would be difficult to take into account that forces acting on the particle must be transmitted by a speed less than that of light, as is demanded by Einstein's theory of special relativity. If the particle were deformable, it would not be truly elementary. Therefore, finite-size particles cannot serve as a good basis for a theory of elementary objects.

In addition, there is an effect that alters the electric charge of the particle. This effect is called "vacuum polarization." During extremely short time intervals, quantum fluctuations cause the creation and subsequent annihilation of particle-antiparticle pairs. If these particles carry electric charges, the charges whose signs are opposite to our particle in question tend to move towards it, and this way they tend to neutralize it. Although this effect is usually quite small, there is a tendency of the vacuum to "screen" the charge of our particle. This screening effect implies that a particle whose charge is Q_{bare} looks like a particle with a smaller charge Q_{phys} when viewed at some distance. The relation between Q_{bare} and Q_{phys} again depends on R, and, as was the case for the mass of the particle, the charge renormalization also tends to infinity as the radius R is sent to zero (even though the effect is usually rather small at finite R).

It was already in the first half of the twentieth century that physicists realized the following. The only properties of a particle such as an electron that we ever measure in an experiment are the physical mass m_{phys} and the physical charge Q_{phys}. So, the procedure we have to apply is that we should take the limit where R is sent to zero while m_{bare} and the bare charge Q_{phys} are kept fixed. Whatever happens to the bare mass m_{bare} and the bare charge Q_{bare} in that limit is irrelevant, since these quantities can never be measured directly.

Of course, there is a danger in this argument. If, in eq. (5), we send R to zero while keeping m_{phys} fixed, we notice that m_{bare} tends to minus infinity. Can theories in which particles have negative mass be nevertheless stable? The answer is no, but fortunately eq. (5) is replaced by a different equation in a quantized theory. m_{bare} tends to zero, not minus infinity.

3 The Renormalization Group

The modern way to discuss the relevance of the small distance structure is by performing scale transformations, using the renormalization group [3, 4], and we can illustrate this again by considering the equation of motion of the planets. Assume that we took definite time intervals Δt, finding equations for the displacements Δx. Imagine that we wish to take the limit $\Delta t \to 0$ very carefully. We may decide first to divide all Δt's and all Δx's by 2 [see Fig. 2(b)]. We observe that, if the original intervals are already sufficiently small, the new results of a calculation will be very nearly the same as the old ones. This is because, during small time intervals, planets and moons move along small sections of their orbits, which are very nearly straight lines, the division by 2 would have made no difference at all. Planets move along straight lines if no force acts on them. The reason why differential equations were at all successful for planets is that we may ignore the effects of the forces (the "interactions") when time and space intervals are taken to be very small.

In quantized field theories for elementary particles, we have learned how to do the same thing. We reconsider the system of interacting particles at very short time and distance scales. If at sufficiently tiny scales the interactions among the particles may be ignored, then we can understand how to take the limits where these scales go all the way to zero. Since then the interactions may be ignored, all particles move undisturbedly at these scales, and so the physics is then understood. Such theories can be based on a sound mathematical footing – we understand how to do calculations by approximating space and time as being divided into finite sections and intervals and taking the limits in the end.

So, what is the situation here? Do the mutual interactions among elementary particles vanish at sufficiently tiny scales? Here is the surprise that physicists had to learn to cope with; they do not.

Many theories indeed show very bad behavior at short distances. A simple proto-type of these is the so-called chiral model [5].

In such a model, a multicomponent scalar field is introduced which obeys a constraint: its total length is assumed to be fixed,

$$\sum_i |\phi_i|^2 = R^2 = \text{fixed} \tag{6}$$

At large distance scales, the effects of this constraint are mild, as the quantum fluctuations are small compared to R. At small distance scales, however, the quantum fluctuations are large compared to R, and hence the nonlinear effects of the constraint are felt much more strongly there. As a consequence, such a theory has large interactions at small distance scales and vice versa. Therefore, at infinitesimally small distance scales, such a theory is ill-defined, and the model is unsuitable for an accurate description of elementary particles. Other examples of models with bad small-distance behavior are the old four-fermion interaction model for the weak interactions and most attempts at making a quantum version of Einstein's gravity theory.

But some specially designed models are not so bad. Examples are: a model with spinless particles whose fields ϕ interact only through a term of the form $\lambda\phi^4$ in the Lagrangian, and a model in which charged particles interact through Maxwell's equations (quantum electrodynamics, QED). In general, we choose the distance scale to be a parameter called $1/\mu$. A scale transformation by a factor of 2 amounts to adding $ln2$ to $ln\mu$, and if the distance scale is Δx, then

$$\frac{\mu d}{d\mu}\Delta x = -\Delta x \tag{7}$$

During the 1960s, it was found that in all theories existing at the time, the interaction parameters, being either the coefficient λ for $\lambda\phi^4$ theory, or the coefficient e^2 in quantum electrodynamics for electrons with charge e, the variation with μ is a positive function [6], called the β function:

$$\frac{\mu d}{d\mu}\lambda = \beta(\lambda) > 0, \tag{8}$$

so, comparing this with eq. (7), λ is seen to increase if Δx decreases.

In the very special models that we just mentioned, the function $\beta(\lambda)$ behaves as λ^2 when λ is small, which is so small that the coupling only varies very slightly as we go from one scale to the next. This implies that, although there are still interactions, no matter how small the scales at which we look, these interactions are not very harmful, and a consequence of this is that

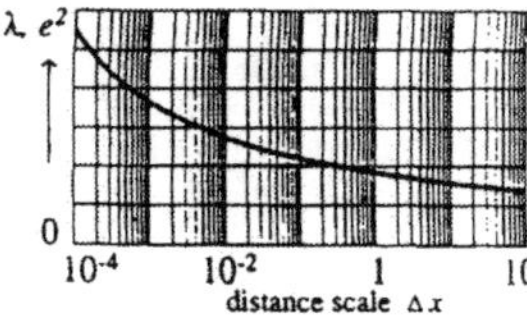
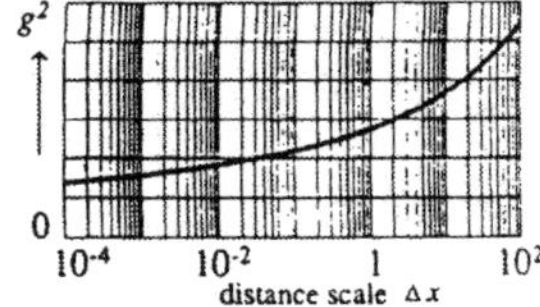

Fig. 3. Scaling of the coupling strength as the distance scale varies (a) for $\lambda\psi^4$ theories and QED, (b) for Yang-Mills theories

these theories are "renormalizable." If we apply the perturbation expansion for small λ then, term by term, the expansion coefficients are uniquely defined, and we might be seduced into believing that there are no real problems with these theories.

However, many experts in these matters were worried indeed, and for good reason: If β is positive, then there will be a scale where the coupling strength among particles diverges. The solution to eq. (8) is [see Fig. 3(a)]

$$\lambda(\mu) = 1/(C - \beta_2 ln\mu), \quad \text{if} \quad \beta(\lambda) = \beta_2\lambda^2, \tag{9}$$

where C is an integration constant, $C = 1/\lambda(1)$ if $\lambda(1)$ is λ measured at the scale $\mu = q$. We see that at scales $\mu = 0[exp(1/\beta_2\lambda(1))]$, the coupling explodes. Since for all $\lambda(1)$ this is exponentially far away, the problem is not noticed in the perturbative formulation of the theory, but it was recognized that if, as in physically realistic theories, λ is taken to be not very small, there is real trouble at some definite scale. And so it was not crazy to conclude that these quantum field theories were sick and that other methods should be sought for describing particle theories.

I was never afflicted with such worries for a very simple reason. Back in 1971, I carried out my own calculations of the scaling properties of field theories, and the first theory I tried was Yang-Mills theory. My finding was, when phrased in modern notation, that for these theories,

$$\beta(g^2) = Cg^4 + 0(g^6) \quad \text{with} \quad C < 0 \tag{10}$$

if the number of fermion species is less than 11 [for SU(2)] or $16\frac{1}{2}$ [for SU(3)]. The calculation, which was alluded to in my first paper on the massive Yang-Mills theory [7], was technically delicate but conceptually not very difficult. I could not possibly imagine what treasure I had here or that none of the experts knew that β could be negative; they had always limited themselves to studying only scalar field theories and quantum electrodynamics, where β is positive.

4 The Standard Model

If we were to confront the infinities in our calculations for the weak-interaction processes, we had to face the challenge of identifying a model for the weak interaction that shows the correct intertwining with the electromagnetic force at large distance scales but is sufficiently weakly interacting at small distances. The resolution here was to make use of spontaneous symmetry breaking. The mass generation mechanism discussed here should, strictly speaking, not be regarded as spontaneous symmetry breaking, since in these theories the vacuum does not break the gauge symmetry. "Hidden symmetry" is a better phrase [8]. We simply refer to this mechanism as the "Higgs mechanism." We use a field with a quartic self-interaction but with a negative mass term, so that its energetically favored value is non vanishing. The fact that such fields can be used to generate massive vector particles was known but not used extensively in the literature. Also the fact that one could construct reasonable models for the weak interaction along these lines was known. These models, however, were thought to be inelegant, and the fact that they were the unique solution to our problems was not realized.

Not only did the newly revived models predict hitherto unknown channels for the weak interaction, they also predicted a new scalar particle, the Higgs boson [9–11]. The new weak interaction, the so-called neutral-current interaction, could be confirmed experimentally within a few years, but as of this writing, the Higgs boson is still fugitive. Some researchers suspect that it does not exist at all. Now if this were true then this would be tantamount to identifying the Higgs field with a chiral field-a field with a fixed length. We could also say that this corresponds to the limiting case in which the Higgs mass was sent to infinity. An infinite-mass particle cannot be produced, so it can be declared to be absent. But as we explained before, chiral theories have bad small-distance behavior. We can also say that the interaction strength at small distances is proportional to the Higgs mass; if that would be taken to be infinite then we would have landed in a situation where the small-distance behavior was out of control. Such models simply do not work. Perhaps experimentalists will not succeed in producing and detecting Higgs particles, but this then would imply that entirely new theories must be found to account for the small-distance structure. Candidates for such theories have been proposed. They seem to be inelegant at present, but of course that could be due to our present limited understanding, who knows? New theories would necessarily imply the existence of many presently unknown particle species, and experimenters would be delighted to detect and study such objects. We cannot lose here. Either the Higgs particle or other particles must be waiting there to be discovered, probably fairly soon [12, 13].

To the strong interactions, the same philosophy applies, but the outcome of our reasoning is very different. The good scaling behavior of pure gauge theories [see Fig. 3(b)] allows us to construct a model in which the interactions at large distance scales is unboundedly strong, yet it decreases to zero (though only logarithmically) at small distances. Such a theory may describe the binding forces between quarks. It was found that these forces obtain a constant strength at arbitrarily large distances, where Coulomb forces would have decreased with an inverse square law. Quantum chromodynamics, a Yang-Mills theory with gauge group SU(3), could therefore serve as a theory for the strong

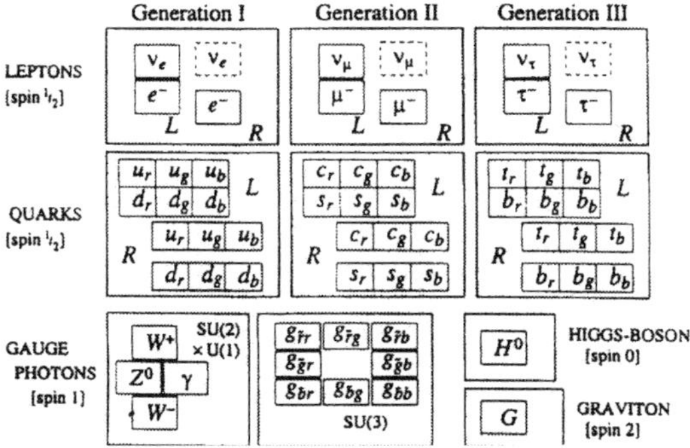

Fig. 4. The standard model

interactions. It is the only allowed model in which the coupling strength is large but nevertheless the small-distance structure is under control.

The weak force, in contrast, decreases exponentially as the distance between weakly interacting objects becomes large. Thus gauge theory allows us to construct models with physically acceptable behavior at short distances, while the forces at large distances may vary in any of the following three distinctive ways:

(i) The force may drop exponentially fast, as in the weak interaction;

(ii) The force may drop according to an inverse square law, as in electromagnetism, or

(iii) The force may tend towards a constant, as in the strong interactions.

The Standard Model is the most accurate model describing nature as it is known today. It is built exactly in accordance with the rules sketched in Fig. 4. Our philosophy is always that the experimentally obtained information about the elementary particles refers to their large-distance behavior. The small-distance structure of the theory is then postulated to be as regular as is possible without violating principles such as strict obedience of casuality and Lorentz invariance. Not only do such models allow us to calculate their implications accurately, it appears that Nature really is built this way. In some sense, this result appears to be too good to be true. We shall shortly explain our reason to suspect the existence of many kinds of particles and forces that could not yet be included in the Standard Model, and that the small-distance structure of the Standard Model does require modification.

5 Future Colliders

Theoreticians are most eager to derive all they want to know about the structures at smaller distances using pure thought and fundamental principles. Unfortunately, our present insights are hopelessly insufficient, and all we have are some wild speculations. Surely, the future of this field still largely depends on the insights to be obtained from new experiments.

The present experiments at the Large Electron Positron Collider (LEP) at CERN are coming to a close. They have provided us with impressive precision measurements that not only gave a beautiful confirmation of the Standard Model, but also allowed us to extrapolate to higher energies, which means that we were allowed a glimpse of structures at the smallest distance ranges yet accessible. The most remarkable result is that the structures there appear to be smooth; new interactions could not be detected, which indicates that the mass of the Higgs particle is not so large, a welcome stimulus for further experimental efforts to detect it.

In the immediate future we may expect interesting new experimental results first from the Tevatron Collider at Fermilab, near Chicago, and then from the Large Hadron Collider (LHC) at CERN, both of which will devote much effort to finding the still elusive Higgs particle. Who will be first depends on what the Higgs mass will turn out to be, as well as other not yet precisely known properties of the Higgs. Detailed analysis of what we know at present indicates that Fermilab has a sizable chance at detecting the Higgs first, and the LHC almost certainly will not only detect these particles, but also measure many of their properties, such as their masses, with high precision. If supersymmetric particles exist, LHC will also be in a good position to be able to detect these, in measurements that are expected to begin shortly after 2005.

These machines, which will discover structures never seen before, however, also have their limits. They stop exactly at the point where our theories become highly interesting, and the need will be felt to proceed further. As before, the options are either to use hadrons such as protons colliding against antiprotons, which has the advantage that, due to their high mass, higher energies can be reached, or alternatively to use leptons, such as e^+ colliding against e^-, which has the advantage that these objects are much more pointlike, and their signals are more suitable for precision experiments [14]. Of course, one should do both. A more ambitious plan is to collide muons, μ^+ against μ^-, since these are leptons with high masses, but this will require numerous technical hurdles to be overcome. Boosting the energies to ever-increasing values requires such machines to be very large. In particular the high-energy electrons will be hard to force into circular orbits, which is why design studies of the future accelerators tend to take the form of straight lines, not circles. These linear accelerators have the interesting feature that they could be extended to larger sizes in the more distant future.

My hope is that efforts and enthusiasm to design and construct such machines in the future will not diminish. As much international cooperation as possible is called for. A sympathetic proposal [15], is called ELOISATRON, a machine in which the highest conceivable energies should be reached in a gigantically large circular tunnel. It could lead to a hundredfold improvement of our spatial resolution. What worries me, however, is that in practice one group, one nation, takes an initiative and then asks other groups and nations to join, not so much in the planning, but rather in financing the whole thing. It is clear to me that the best international collaborations arise when all partners are involved from the very earliest stages of the development onwards. The best successes will come from those institutions that are the closest approximations to what could be called "world machines." CERN claims to be a world machine, and indeed as such this laboratory has been, and hopefully will continue to be, extremely successful. Unfortunately, it still has an E in its name. This E should be made as meaningless as the N (after all, the physics studied at CERN has long ago ceased to be nuclear, it is subnuclear now). I would not propose to change the name, but to keep the name CERN only to commemorate its rich history.

6 Beyond the Standard Model

Other, equally interesting large scientific enterprises will be multinational by their very nature: plans are underway to construct neutrino beams that go right through the earth to be detected at the exit point, where it may be established how subtle oscillations due to their small mass values may have caused transitions from one type into another. Making world machines will not imply that competition will be eliminated; the competition, however, will not be between nations, but rather between the different collaborators who use different machines and different approaches towards physics questions.

The most interesting and important experiments are those of which we cannot guess the outcome reliably. This is exactly the case for the LHC experiments that are planned for the near future. What we do know is that the Standard Model, as it stands today, cannot be entirely correct, in spite of the fact that the interactions stay weak at ultrashort distance scales. Weakness of the interactions at short distances is not enough; we also insist that there be a certain amount of stability. Let us use the metaphor of the planets in their orbits once again. We insisted that, during extremely short time intervals, the effects of the forces acting on the planets have hardly any effect on their velocities, so that they move approximately in straight lines. In our present theories, it is as if at short time intervals several extremely strong forces act on the planets, but, for some reason, they all but balance out.

It is generally agreed that the most attractive scenario is one involving "supersymmetry," a symmetry relating fermionic particles, whose spin is an integer plus one-half, and bosonic particles, which have integral spin. (Supersymmetry has a vast literature. See, for instance, the collection of papers in [16, 17]). It is the only symmetry that can be made to do the required job in the presence of the scalar fields that provide the Higgs mechanism, in an environment where all elementary particles interact weakly. However, when the interactions do eventually become strong then there are other scenarios. In that case, the objects playing the role of Higgs particles may be not elementary objects but composites, similar to the so-called Cooper pairs of bound electrons that perform a Higgs mechanism in ultra cool solid substances, leading to superconductivity. Just because such phenomena are well known in physics, this is a scenario that cannot easily be dismissed. But since there is no evidence at present of a new strong interaction domain at the TeV scale, the bound-state Higgs theory is not favored by most investigators.

One of the problems with the supersymmetry scenario is the supersymmetry breaking mechanism. Since at the distance scale where experiments are done at present no supersymmetry has been detected, the symmetry is broken. It is assumed that the breaking is "soft," which means that its effects are seen only at large distances, and only at the tiniest possible distance scales is the symmetry realized. Mathematically, this is a possibility, but there is as yet no plausible physical explanation of scales, where the gravitational force comes into play.

Until the early 1980s the most promising model for the gravitational force was a supersymmetric variety of gravity: supergravity [17]. It appeared that the infinities that were insurmountable in a plain gravity theory would be overcome in supergravity. Curiously, however, the infinities appeared to be controlled by the enhanced symmetry and not by an improved small-distance structure of the theory. Newton's constant, even if controlled by a dilaton field, still is dimensionful in such theories, with consequently uncontrolled strong interactions in the small-distance domain. As the small-distance structure of the theory was not understood, it appeared to be almost impossible to draw conclusions from the theory that could shed further light on empirical features of our world.

An era followed with even wilder speculations concerning the nature of the gravitational force. By far the most popular and potentially powerful theory is that of the superstrings [18]. The theory started out by presenting particles as made up of (either closed or open) pieces of string. Fermions living on the string provide it with a supersymmetric pattern, which may be the origin of the approximate pattern, which may be the origin of the approximate supersymmetry that we need in our theories. It is now understood that only in a perturbative formulation do particles look like strings. In a non perturbative formalism there seems to be a need not only of strings but also of higher dimensional substances such as membranes. But what exactly is the perturbation expansion in question? It is not the approximation that can be used at the shortest infinitesimal distances. Instead, the shortest distances seem to be linked to the largest distances by means of duality relations. Just because superstrings are also held responsible for the gravitational force, they cause curvature of space and time to such an extent that it appears to be futile to consider distances short compared to the Planck scale.

According to superstring theory, it is a natural and inevitable aspect of the theory that distance scales shorter than the Planck scale cannot be properly addressed, and we should not worry about it. When outsiders or sometimes colleagues from unrelated branches of physics attack superstring theory, I come to its defense. The ideas are very powerful and promising. But when among friends, I have this critical note. As string theory makes heavy use of differential equations it is clear that some sort of continuity is counted on. We should attempt to find an improved short-distance formulation of theories of this sort, if only to justify the use of differential equations or even functional integrals.

Rather than regarding the above as criticism against existing theories, one should take our observations as indications of where to search for further improvements. Emphasizing the flaws of the existing constructions is the best way to find new and improved procedures. Only in this way can we hope to achieve theories that allow us to explain the observed structures of the Standard Model and to arrive at more new predictions, so that we can tell our experimental friends where to search for new particles and forces.

References

1. R.P. Crease and C.C. Mann, *The Second Creation: Makers of the Revolution in Twentieth-century Physics*, Macmillan, New York (1986).
2. A. Pais, *Inward Bound: Of Matter and Forces in the Physical World*, Oxford University, London (1986).
3. K.G. Wilson and J. Kogut, Phys. Rep., Phys. Lett. C **12**, 75 (1974).
4. H.D. Politzer, Phys. Rep., Phys. Lett. C **14**, 129 (1974).
5. B.W. Lee, *Chiral Dynamics*, Gordon and Breach, New York (1972), pp. 60–67.
6. D.J. Gross, in *The Rise of the Standard Model*, Cambridge University, Cambridge (1997), p. 199.
7. G't Hooft, Nucl. Phys. B **35**, 167 (1971).
8. S. Coleman, "Secret Symmetries", in *Laws of Hadronic Matter*, ed. A. Zichichi, Academic, New York, London (1975).
9. P.W. Higgs, Phys. Lett. 12, 132 (1964a).
10. P.W. Higgs, Phys. Rev. Lett. **13**, 321 (1964b).
11. P.W. Higgs, Phys. Rev. **145**, 1156 (1966).
12. E. Accomando, et al., Phys. Rep. **299**, 1 (1998).
13. P.M. Zerwas, "Physics with an e^+e^- linear collider at high luminosity", Cargese lectures 1999, prepring DESY, 99–178.
14. J. Ellis, *Possible Accelerators at CERN beyond the LHC*, preprint CERN-TH/99-350, hep-ph/9911440 (1999).
15. A. Zichichi, "Fifty years of subnuclear physics: From past to future and the ELN project", in *Highlights of Subnuclear Physics: 50 Years Later: Proceedings of the International School of Subnuclear Physics*, ed. A. Zichichi, World Scientific, Singapore and River Edge, London (1999), p. 161.
16. S. Ferrara, Ed. *Supersymmetry*, Vol. 1, North Holland, Amsterdam (1987).
17. S. Ferrara, Ed. *Supersymmetry*, Vol. 2, North Holland, Amsterdam (1987).
18. J. Polchinski, *String Theory*, Vol. 1, *An Introduction to the Bosonic String, Cambridge Monographs on Mathematical Physics*, ed. P.V. Landshoff et al., Cambridge University, Cambridge (1998).

The Creative and Unpredictable Interaction of Science and Technology

Charles Townes

University of California, Berkeley, U.S.A

Fig. 1. Charles Townes after receiving the B.M. Birla Science Centre's Life Time Achievement in Science Award

Charles Hard Townes was born in Greenville, South Carolina on 28 July, 1915. His father was an attorney. After schooling, Charles joined the Furman University, Greenville where he received the Bachelor of Science degree in Physics and Bachelor of Arts degree in Modern Languages in 1935. From his early days Charles was fascinated by Physics. After completing his Masters degree in Physics at Duke University in 1936, he joined the California Institute of Technology for his PhD, which he received in 1939 for his Thesis on isotope seperation and nuclear spins.

From 1933 to 1947 Dr. Townes worked at the Bell Telephone Laboratories. During the war period he worked on designing radar controlled bomb targeting system and obtained a number of relevant patents. After the war he worked on using the war time microwave radar research to spectroscopy. His belief was that this was a powerful technique for studying the structure of molecules and atoms and it could even control the electromagnetic waves. Meanwhile in 1941

B.G. Sidharth (ed.), *A Century of Ideas*, 123–137.
© Springer Science+Business Media B.V. 2008

Dr. Townes was married to Frances H Brown. They had four daughters. He was appointed as a faculty member at Columbia University in 1948. Here he continued doing research in microwave physics, particularly the interactions between microwaves and molecules. Dr. Townes got the idea of MASER in 1951. He and his coworkers began working on a device that used ammonia as the medium. By 1954 they had achieved success and could amplify and generate electromagnetic waves by stimulated emission. He and his students named the device a MASER, an acronym for Microwave Amplification by Stimulated Emission of Radiation.

In 1958 Dr. Townes and his brother-in-law, Dr. Schavlow of Stanford showed that MASERS could be made to operate in the optical and infra red region. This was the birth of LASERS, or Light Amplification of Stimulated Emission of Radiation.

In 1964 he was awarded the Nobel Prize for Physics jointly with the Russian team of A.M. Prokhorov and N.G. Basov of the Lebedev Institute, for their independent and "fundamental work in the field of quantum electronics, which has led to the construction of oscillators and amplifiers based on the maser-laser principle."

Today both LASERS and MASERS have any number of applications from astronomy, through communications to medicine and industry. From 1959 to 1961, Prof. Townes was on leave of absence from Columbia University and served as Director Research of the Institute for Defence Analyses in Washington DC. In 1961 Prof. Townes was appointed Provost and Professor of Physics at the Massachusetts Institute of Technology. In 1966 he became the Institute Professor at MIT. The same year he relinquished his position as Provost in order to return to more intense research in the fields of quantum electronics and astronomy. In 1967 he was appointed Professor at the University of California. He has remained there since.

Prof. Townes has received a large number of honors, awards, fellowships and degrees. These include the Guggenheim Fellowship, and the Fulbright Lecturership. He has also served on the scientific advisory board of the US Airforce, was Chairman of the Strategic Weapons Panel of the Department of Defence, he was Chairman of the Science and Technology Advisory Committee for Manned Space Flight of NASA and a Member of the President's Science Advisory Committee and so on. His honorary degrees include the D.Litt from Furman University, the ScD of Clemson College, the ScD of Columbia University, the ScD of Duke University and so on. His awards and honors include the Research Corporation Annual Award, the IRE Morris N. Liebmann Memorial Prize, the Comstock Award, National Academy of Sciences, the Exceptional Service Award, U.S. Air Force; the Thomas Young Medal and Prize of the Institute of Physics and the Physical Society of England, the IEEE Medal of Honor and the B.M. Birla Science Centre's Life Time Achievement in Science Award.

Though Prof. Townes was involved with Airforce and Defence establishments, he is a very humane person who stresses the need for equitable and harmonious relationships between various countries. His point is that unless there is equity and harmony amongst nations and within societies, there will not be peace and without peace, the wonderful fruits of science and technology cannot be enjoyed by mankind. This apart he is very open to new ideas and is not overawed by objections. Age has not diminished his mental or physical agility.

I am very pleased to have this opportunity to visit Hyderabad and to give this year's B.M. Birla Memorial Lecture. We will discuss the nature of interactions between science and technology and between scientists as this affects new science and technology.

Frequently people believe that science is somehow primarily created by lone scientists thinking hard, which creates new science and ideas. There is a little truth in that. But especially today, the rapid growth of science and of technology depends a great deal on the interaction between people, the trading of their personal ideas, and interdisciplinary interactions. What I would call the sociology of science and technology is very important to their rapid and successful growth. Another aspect which is very important is a sense of openness and willingness to explore. We cannot predict what's going to be discovered in science. The new things are new. We can foresee some things, or some developments. But discovery always leads to enormous surprises. We have to be very open, and encourage new ideas. We must encourage young people in new approaches, and encourage exploration. It is characteristically unexpected areas and exploration which have really transformed our society. The most striking products of science and technology have helped humans enormously.

It's probably useful to illustrate unpredictability and surprise by a few examples. Let's consider firstly something in the past – our aircraft. In 1895 Lord Kelvin, who was one of the great scientists of the day, said "Heavier than air flying machines are impossible". Then there was Lord Rayleigh, who said "I have not the smallest molecule of faith in aerial navigation other than ballooning". These were the two most important physicists of the day. What happened? Seven years later the Wright Brothers were flying an aircraft. We know today how much we enjoy and depend on aircraft heavier than air, and how obvious it is to everyone that they can work. A few decades later, in 1933, Lord Rutherford said "Anyone who expects a source of power from transformation of these atoms is talking moonshine". He said that publicly because some people were saying mass has energy according to relativity theory, and one should get some energy from it somehow, perhaps from radioactive nuclei. Lord Rutherford's statement was affirmed publicly by additional prominent physicists of that time so people wouldn't be fooling themselves about getting energy out of nuclei. But it was only six years later that fusion was discovered. I was a graduate student at that time, and remember that many students and faculty were suddenly saying "Look, Fermi's findings were in error and led him to think that he was making heavier elements when in fact he observed fusion". The truth was discovered by a chemist, not a physicist. And people began to realize that yes, we could indeed get a great deal of energy from atoms.

In 1937 there was a commission formed by Franklin Roosevelt, President of the United States, the so-called Roosevelt Commission whose assignment was to advise on what would be the most important technical developments over the next few decades, how they might effect the United States, and hence what technical developments the government should sponsor and emphasize. This commission was made of distinguished scientists and engineers. They thought hard and made a report. In it they mentioned improvements of agriculture. Certainly improvements in agriculture can occur and are helpful. The report also emphasized the importance of improving the efficiency of machinery. It mentioned a number of useful things. But the changes which are most important are what they missed. This report was in 1937. What did they miss? They missed nuclear energy, which came along only a couple of years later. They missed radar, which became prominent only a few years later. They missed the transistor and solid state electronics. They of course missed the laser which also came about within the time scale of their attempted foresight. Other things missed were magnetic resonance, jet aircraft, rocketry, space travel, antibiotics and so on. They missed a large number of important things, but this was not an ignorant group of people. They were senior scientists and engineers who were responsible to try to predict what we should be looking for and developing.

If we go forward a little in time there are other cases. In 1956, Richard Woolley, Royal Astronomer of Great Britain, pronounced publicly that "space travel is utter bilge". One year later Sputnik went up. Suddenly people had a different view, and of course in 1969, only about a decade later, we were landing on the moon.

We are now enjoying the computer age. Ken Olsen, the president and founder of Digital Equipment Company, and a person who should have been very knowledgeable, said in about 1980 "There is no reason anyone would want a computer in their home." Well, many of us don't agree with that now. But that was his expectation of the industry. In 1981, Bill Gates, founder of Microsoft and for whom we of course have great admiration, said of information units "640 K ought to be enough for anyone". We are now up in the billions and more.

The simple illustrations mentioned show how difficult it is for us to predict the new things that are going to be discovered. What we must do is to be open, we must search, we must explore, and creative people must interact and trade ideas.

Another common but somewhat distorted idea is that the way technology develops is through discoveries in science. New scientific ideas and principles get applied to technical things and produce technology. So we go from science to technology that is useful to humans. There is indeed some truth to that. But I don't think that the public generally realizes that science depends very much on technology just as technology depends very much on science. My point of view is that outstanding science and technology must go together, interact strongly, and must both be present within a community if they are going to be good. Furthermore technology contributes very importantly to science; I will give some illustrations. Consider, for example, the study of noise by engineers for communication purposes. Well, that's a kind of dirty subject. Not many scientists would be interested in how much noise there is around, but for communications engineers it is of course important, so they examine it carefully.

J.B. Johnson at the Bell Telephone Laboratories examined the noise fluctuations in circuitry. He found that there was a noise voltage proportional to the resistance in the circuit. That was a striking discovery, and an applied mathematician at Bell Laboratories then showed that is a very fundamental result of thermodynamics. And it is now called Johnson noise. Another examination of noise, back in the 1930s, was done by Karl Jansky. He was asked to look for what radio or microwave noise might be picked up by antennas, and so constructed a good antenna and looked around with it. He found there was radio noise coming from somewhere in the sky, and with astronomers he was able to show that it was the centre of the Milky Way which was producing a lot of noise. That was the origin of radio astronomy, a very important scientific field these days. Just a good engineering examination of noise produced a new field of astronomy. The next thing that happened in examination of noise, again at Bell Telephone Laboratories, was that Arno Penzias and Bob Wilson looked more carefully at the distribution of noise coming from all parts of the sky. They did a very responsible job and I am very proud that Arno had been one of my students at Columbia University. He and Wilson looked carefully and found that there was noise coming from all directions. Not very strong noise, but they had sensitivity enough using a maser to detect rather weak noise coming from all directions. What did that mean? It was the detection of the origin of our Universe – the remnants of the Big Bang which occurred roughly 14 billion years ago. What could be more fundamental in science than discovering the origin of our universe by studying noise, of all things?

There are many other illustrations of the importance of engineering for science, but I would mention only one more rather general thing of importance. During World War II we scientists, many in the United States, Europe, and other places, had to pay attention to engineering in trying to help the military. In the United States scientists and engineers did a great deal of work on radar as well as nuclear energy. I was pulled into the radar business and learned a lot about microwaves. Many other scientists learned engineering that they didn't know before. Engineers and scientists were not terribly close before that, but the War brought engineering and science together. And much of my own science, including microwave spectroscopy, the maser and the laser, grew out of my engineering experience.

In addition, other physicists went back to universities to work again in basic science. They had seen new possibilities from their engineering work and out of that came new things like nuclear magnetic resonance, discovery of the radio waves from hydrogen atoms in other space, development of radio astronomy, and microwave spectroscopy, my own field. I had recognized from studying transmission of radar waves through the atmosphere that we could do very precise work of molecules, atoms, and nuclei with microwave spectroscopy.

How does industry and how does our society decide what research to sponsor in order to have a good development of technology? One can find at least some cases where good decisions have been made. One case is that of Mervin Kelly who was director of research at Bell Telephone Laboratories in the 1930s. He recognized that solid state physics was beginning to be understood. During the 1930s the theory of solid state physics was being developed and we could understand better how the electrons moved around in conductors or semiconductors. Work at Bell Telephone Laboratories dealt with solids, resistors, conductors, and so on. Kelly decided that solid state physics might in the long run help in understanding how best to produce and use solids. So he hired a few solid state physicists in the late 1930s and a few more after World War II. Actually, many people think that the transistor grew out of Kelly's idea that he might be able to get a solid state amplifier and that's why he hired these people. But that's not the case at all. He hired them simply because Bell Labs was dealing with materials and the basic science field was beginning to develop. The transistor was discovered by accident. Walter Brattain was measuring surface properties of a solid by putting a contact on a semiconductor. He found a peculiar behavior which he couldn't understand. Since he couldn't figure out what was going on he got the help of John Bardeen, one of the theoretical solid state physicists recently hired by Bell Telephone Laboratories. Bardeen studied Brattain's results, thought about them, and then said "Hey, that's amplification". And he figured out how this was occurring – this was the discovery of the first transistor. William Shockley was another theorist at Bell Telephone Laboratory. He was abroad at that time. He came back very quickly and was very excited about the result. So he worked on the new phenomena and invented additional types of transistors. That's how the transistor came into being. An accidental discovery – yes, but also thoughtful planning on the part of the administration of the Bell Laboratories.

Now let me give a case where even Bell Laboratories seems to have gone in the wrong direction. The laser came out of microwave spectroscopy. I can illustrate that a little bit later in a way which I believe will convince you. Now what research director wanting a bright light, brighter than any presently available, would hire people to work on the microwave spectra of molecules? Instead, he would probably go to a company making lights and say please make me something brighter. And they might make it perhaps twice as bright. But the laser gives us billions of times more intensity that we have ever had before. And no director would have recognized that the way to go was to study molecules with microwaves. But it was. I was doing this type of work because I thought it was good science, and Bell Laboratories would allow me to do it. The same field grew up very quickly, immediately after the war, in other companies which had radar microwave equipment. Friends of mine at General Electric, at RCA, and at Westinghouse were also doing microwave spectroscopy and they did it very well. But those companies eventually rejected the field and told my friends "sorry, we just can't support you doing this kind of work. You have to do something for us, something with useful applications", and so they had to stop. Bell Laboratories wanted me to do some kind of engineering that would be useful rather than microwave spectroscopy. But I said no I really wanted to do this and they allowed me to do it. They wouldn't hire anyone to help me but they did allow me to do it. That was generous when they didn't expect anything very useful to come from it. Well, the field became very interesting scientifically and I was offered a professorship at Columbia University, so I moved to a university.

For extending my field of work, I wanted to get to shorter wavelengths. Microwaves at that time could have wavelengths as short as about half a centimeter, and that was giving us a lot of good results. But I knew if I could get shorter wavelengths I would see more things and additional interesting science. I wanted to produce waves shorter than a millimeter, and on down into the far infrared region. How could one do that? We were working with cyclotrons and magnetrons at that time. They simply couldn't be made small enough to operate and produce those wavelengths. I tried many things and my students tried various experiments to produce short waves. I had various ideas, like sending beams of electrons along special types of surfaces, or producing harmonics. Some of my ideas worked a bit but not well enough. For several years I kept trying to get to shorter wavelengths.

A national committee was formed to examine how we can get to shorter wavelengths, and I was appointed chairman. It was a big committee and we travelled a lot to see what ideas anyone else might have. We went to England, France and Germany, as well as covering the United States. We didn't find any great ideas. So I was getting discouraged. We were to have our last meeting in Washington, DC. Before the meeting I woke up early one morning worrying over this and went out and sat in a park on a nice bright day, with flowers blooming beautifully. I said to myself, "Why haven't we been able to do this?" I went through in my head the various things we had thought of. I had thought of possibly using molecules to generate these waves. Molecules can oscillate very fast, producing frequencies in the infrared. They can generate electromagnetic waves, but I was proud to recognize that the second law of thermodynamics meant that they couldn't produce very much power. I was locked into the laws of thermodynamics. In my mind I thought, yes, molecules can produce these frequencies but we can't get much power from them because of thermodynamics. Then I suddenly gasped. Wait a minute, molecules do not have to be at a defined temperature and obey thermodynamics. If we get molecules only in excited energy states rather than lower energy states, then we could get real intensity.

Let me now just clarify the physics involved in masers and lasers. What a laser involves is the following: Take a molecule or atom which can be in a low energy state or in a high energy state. If it is at high energy it can spontaneously drop down to low energy and make a photon, giving light or radiation of some kind. If it is in a low energy state and a photon comes along, the photon's energy can excite the molecule and energy of the light wave decreases. But if it is in a high energy state and a wave comes along the wave interacts with the molecule, it falls down and gives its energy to the wave, thus building up energy in the wave. The trouble with ordinary molecules is that more of them are in the low energy state than the high energy state. So more of the wave gets absorbed by low energy states and then that is given energy by the high energy states. All we have to do is to have many more molecules in the high than in the low energy state. Then as the wave interacts with the molecules, it gets energy from them. How could I arrange that? Well, I had just recently heard an interesting lecture by the German physicist Wolfgang Paul who discussed using beams of molecules deflected by electric force. He had four rods with charges on them, producing a strong electric field. Some molecules would be deflected and other molecules would not be deflected. There was then a way of sending a beam of molecules along the rods, letting the low energy molecules get deflected and thrown away and having the high energy molecules go straight ahead. That was a possible idea. I pulled out a paper and pen to calculate whether it would work. Could I indeed get enough molecules? The molecules could be sent into cavity, a cavity which would resonate with waves going back and forth as the molecules came in, and extracting energy from them. My rough calculation made the idea look very promising. My calculation showed that "Hey that might give a new kind of oscillator!"

I went back to Columbia University and in a few months had persuaded a very good student, James Gordon, to try to do this as his thesis. I wanted to try to make this new type of oscillation first in the microwave region because I had microwave equipment on hand, but then eventually to go on to infrared wavelengths. So we were building a microwave system to try out the idea, using World War II microwave equipment. We worked on it hard. Nobody at Columbia or elsewhere seemed really interested. People would come into my laboratory, I would explain what we were doing, but nobody else wanted to try. For two years we worked on it and hadn't succeeded. Well, at that point the head of the department, Professor Polycarp Kusch, and the former head of the department, Professor I.I. Rabi, both of whom have won Nobel prizes, came into my office, sat down, and said, "Look Charlie, that's not going to work. We know it's not going to work. You know it's not going to work. You have got to stop. You are wasting the University's money". Well, I had looked through the quantum mechanics theory involved very carefully, calculated the appropriate numbers, and I thought it had a good chance of working. Also, I had tenure. You see, tenure in a University means you can't be fired because somebody doesn't agree with you. So I simply told them I thought it had a good chance of working and I was going to continue. They were not pleased. Three months later, the student working on this project dashed into my classroom and said "it's working". We all left the classroom and went to see this new kind of oscillator.

The successful operation of the first maser was of course very exciting to me. So it was also for many people and the newspapers played up this new kind of oscillator. Professor Kusch even apologized to me and said "Well, I guess I should have known that you know more about what you are doing than I do". Kusch and Rabi weren't trying to injure me. They just thought I was wrong. But nevertheless once it worked they were pleased about it. It was so exciting that a number of other physicists started doing this type of work.

Shortly after the maser was working I was scheduled for a sabbatical leave. I thought it was good time to go on a sabbatical, travel around and see what other scientists were doing, and get some fresh ideas. We went to Paris, and there I ran into a former student of mine who was working there at the time and I asked him what he was doing. He said "Oh, I am working on electron spins in magnetic fields and we have found a crystal where an electron spin stays in one direction for a long time". If an electron spin is up in a magnetic field, there is energy. If it is down it has less energy. So an electron changing from pointing upwards to pointing downwards represents something like a molecule making a transition. And I said "Oh, that's just the thing I have been wanting. I want to make a tunable maser that we can vary in frequency, and varying the magnetic field will do that. You vary the magnetic energy and thus vary the frequency. If we can make an electron point upward for a long time that's just what we need and we can make a tunable maser". So with my former student and a French scientist we worked on trying to build this new kind of maser. We got only a little amplification with this new kind of maser before I had to leave Paris. In the meantime, Professor Strandberg, at the Massachusetts Institute of Technology had apparently heard about this. He gave a talk at MIT about the possibility of making a maser with electron spins in a solid. Professor Bloembergen from Harvard was there and sitting in the back of the room. After the talk he said "Well ok, but why would anybody want to do that". "Oh", said Strandberg, "This can make the most sensitive amplifier we have ever had". Bloembergen had been working on electron spins in crystals. He was familiar with this kind of physics and very soon invented the so-called three level solid state maser. It became a very important amplifier, and as an important maser it created quite a stir. In fact many scientists were excited at the maser by that time and they sent in so many papers about it to the Physical Review that the Physical Review editor said "I'm not going to take any more papers on masers – we have had more than we can stand. We have got to have room for other things". Now by then, of course, industrial companies had woken up to the new possibilities provided by masers. They had hired students of mine and others to work on it and help develop the field.

I myself still wanted to get to shorter wavelengths. How to do it? My apologies for mentioning these details which are all kind of personal history, but one has to look at the details to see and understand the importance of interaction between people and how that helps a field develop. Well, I wanted to get to shorter wavelengths. Essentially nobody thought that masers could be made to get to much shorter wavelengths than about one millimeter, but I still wanted to.

I might mention here the origin of the name maser. Well, I had lunch with my students and said we need to find a simple name for the new device after it had been made to work. And so over lunch we invented the word MASER for Microwave Amplification by Stimulated Emission Radiation. Now laser uses exactly the same principles. It gives Light Amplification by Stimulated Emission Radiation. The only difference is the wavelength. We generally say that microwaves go down a wavelength of 1 millimeter, and that shorter than one millimeter there is infrared radiation, which is light. So systems producing waves shorter than one millimeter are lasers, those longer than one millimeter are masers. My students even suggested that maybe we should use the name "iraser" for infra red amplification, but that name didn't last.

I wanted to get to shorter wavelengths, at least infrared radiation, but I hadn't had a very good idea just how to do it. I finally sat down at my desk and said, I am just going to try to push this as far as I can, and see how far we can go. I was sure we could get below one millimeter and I started writing down equations for what was needed and what might be done. From this, I suddenly realized "wait a minute; it is not going to be so hard. We can get right on down to light waves". Generally physicists had all thought that one couldn't produce light waves because for these short wavelengths molecules fall down so much faster from the upper level to the lower level that very few molecules could be kept in upper energy states. Well, I wrote down the equations, and after looking at the numbers realized that there is no reason why we couldn't get right on down to light waves. It was very exciting.

At that time I was consulting at Bell Telephone Laboratories and my job was to talk to people about what they were doing and encourage them. So as part of my consulting at Bell Labs I talked with Art Schawlow, my brother in law. My younger sister had married him while he was working with me as a post doc at Columbia, which pleased me because he was an outstanding scientist and a nice person. I told Art Schawlow that it looked to me like we can make masers go down even to light waves. "Oh" he said, "I have been wondering about that". He seemed very interested and suggested that we work on it together. And he produced an important part of the idea which I had missed. I saw how to excite atoms and molecules and have them oscillate and produce light waves, but I couldn't think of a good resonant cavity.

I recognized that we could make a box and the waves could resonate back and forth, which would work but it wasn't very ideal. Schawlow had the idea of two parallel mirrors, which provided very simple and pure resonant modes. We wrote a paper about how an optical maser, or laser, might be made and what it could do, and that started off the field.

I want now to show you the common skepticism to new ideas, and how important it is for our minds to be open. Our ideas occurred in 1957–1958. It was in the fall of 1957 that I had sat at my desk and had the idea. Earlier in 1957 I was on a committee for the Air Force to try to predict some of the future technology of importance to the Air Force. I was on the electronics committee which wrote a report in the summer of 1957. I put into it that we should push on masers, improve them, and try to push them down to wavelengths at least as short as the mid infrared. That was in our report. The Air Force liked the report but decided to make a further study the following summer. I decided not to serve on the committee the second summer in 1958. So in 1958 the final report was issued. The committee had eliminated what I had said about getting into the infrared because nobody believed it. The committee felt it was just another funny idea of Townes and the maser couldn't get to the infrared. That summer, Schawlow and I had already written the paper on how to do it. They didn't know of our paper even though it was available in unpublished form.

By the time people had seen our paper in the late summer of 1958 there was a lot of excitement. A number of physicists began working on masers for short wavelengths (or lasers), and it is important to recognize that essentially all of the new lasers were produced in industry. They were produced in industrial labs by the students industry had hired from the universities which had been working in the fields of microwave and radio spectroscopy. Industry had by then recognized the importance of the maser, and hired students familiar with the field. Furthermore, someone in industry can spend time working very intensively on something once it becomes exciting, and industry would support it. So the first laser was built by Ted Maiman at the Hughes Laboratory. He had been a student of Professor Willis Lamb and had worked in microwave and radio spectroscopy. The second laser was built at the IBM laboratories by a student of mine, Mirek Stevenson, and a student of Bloembergen, Peter Sorokin. The third type of laser was produced at Bell Telephone Laboratories by Ali Javan, a student of mine (by the way, he was from Iran). Ali Javan had gone to Bell Laboratories and he produced this third kind of laser, the He-Ne discharge laser which has become a very important one. So the laser was produced in industry because industry had learnt of the importance of the field and hired the right young scientists – ones who came out of this field of microwave and radio spectroscopy which industry previously had not considered of value to them. The fourth type of laser, a solid state one, was built at General Electric by Hall and others. These were solid state physicists because they knew solid state materials and recognized how to do it with semi-conductors.

Now again I want to emphasize the interaction between people and the trading of ideas between people. One might say that there were basically no really new ideas in the laser or the maser. Why didn't we make them before? It could in fact have occurred several decades earlier. There was no single idea involved that had not been known by somebody long before the laser came about. Einstein first recognized that one could get energy from molecules by stimulation. That was in 1917. People thought about and worked on the idea. Spectroscopy was a popular field of physics and physicists were very familiar with the interaction of molecules and waves. And quantum mechanics was studied hard and developed during the first half of this century. The problem was that engineers knew about resonators and oscillators. But they didn't know quantum mechanics well at that time. Physicists knew quantum mechanics. But they didn't know so much about oscillators or resonators nor were they very interested. It was the combination of ideas and the recognition of the importance and possibility of producing short wavelengths that were new. These ideas could have come together earlier and lasers might have been made several decades earlier. It was probably because of my interests and my background in engineering plus my knowledge of quantum mechanics that the right idea occurred to me. I have said the whole field came out of microwave spectroscopy. Why do I say that? Well of course I myself worked in microwave spectroscopy. My primary goal was to get a new scientific instrument by producing very pure waves down into the infrared and maybe further. Other people working in this same field had very similar ideas. This includes Nicolai Basov and Alexander Prokhorov in Russia, who of course got the Nobel Prize with me for this discovery. They had been working in microwave spectroscopy when they had the idea. The third independent idea was that of Joe Weber of the University of Maryland. Jose Weber also had a somewhat similar idea though he didn't push it very far. He also had been working in microwave spectroscopy. That's why one can say it came out of that field, which was a combination of quantum mechanics and spectroscopy, and associated strongly with electrical engineering through its dependence on microwaves.

Now what has the laser done? Many friends said to me, immediately after we had the laser idea, "well, it's a nice idea but what kind of application can it have? It's a solution looking for a problem". I knew there were many applications because it married optics and electronics, both of which have many applications. But I could immediately see only a few of the myriad of applications which have developed. We even took the idea to the patent department at Bell Laboratories and a patent lawyer there said "well, I don't think we want to patent it because light waves have not been useful for communications and probably are not of importance to us". Well, Schawlow and I knew of course that couldn't be right. The lawyer said that if you can show us how this new device can do communications, then okay, we will patent it. That was easy to do and we did it. We could see many applications in communications. But consider now some of the other applications. First, the power a laser produces can be very high – so far the power has gotten up to a million billion watts. That's more than all the power used by humans on Earth. That power can be focused to a very small point, much smaller than you can see. Think of the power concentration! Actually it gives us a new state of matter, and might be used to study or produce nuclear fusion. Laser radiation can also be very delicate and precise. Laser tweezers can pick up a single microorganism or molecule, move it from one place to another, and put it down in the right place. Biologists use this in experiments without injuring the cell or microorganism. Lasers give us our present standard of length. We have now redefined the standard of length by using lasers. They give us atomic clocks which are highly precise. For example, the GPS system, which can locate everything on earth, depends on a hydrogen maser acting as an atomic clock. To me the most emotionally pleasing applications are medical ones, particularly when lasers help people with their eyes.

But as an illustration of our difficulty in foreseeing new things, here is another example. I tried to write a paper on the medical applications of lasers with a doctor who wanted to do it. We wrote a paper, but never mentioned the possible application of reattaching a detached retina. I had never heard of a detached retina. How could I foresee that? There have now been about a dozen or more Nobel prizes given to people who have used masers or lasers as a tool. They are important scientific tools in many fields, and that particularly pleases me. I am now using lasers in astronomy. Lasers provide a new way of doing astronomy, and I use them for my own work in astronomy. But communications is perhaps the most important commercial application, along with data recording. Lasers are strikingly useful in a large number of fields. I think the field of lasers is now just in an adolescent stage. We see it as a very powerful field. It can do a lot of things. But it still has a long distance to go.

I feel somewhat apologetic for mentioning so many personal events, but I do that because I think one needs to understand the importance of interactions between people and between fields. One more example is perhaps useful. On my sabbatical I went from Paris to Tokyo and spent some time at the University of Tokyo. As I was walking along a street in Tokyo somebody came up and I recognized another professor from Columbia University, a biologist. I naturally said "hello, what are you doing here". "Oh, I am on sabbatical" he said "Well, what are you doing?" "I am studying a paper written by a theorist in Great Britain named Coulson, who has examined the fluctuations of microbial colonies". A microbe can of course split and that makes two. That's very much like a photon coming along and striking a molecule, making another photon. It can make two. And a microbe can die. That's very much like the molecule absorbing a photon when it comes along. Now what Coulson has done is to allow certain probabilities of splitting and multiplying and of dying, and produced the right mathematics to show the fluctuations in population of a microbial colony. It's a very interesting paper. I said "Wait a minute. That's just what I need. I have been trying to figure out the noise fluctuations in a maser amplifier, and that's just what I need". I must add one more term, that is spontaneous emission where the molecules can spontaneously produce a photon, like the spontaneous creation of a microorganism (which of course can't actually occur). I asked where I could find Coulson's mathematical paper. So I got the paper, studied it, and tried to work out equations for maser noise fluctuations. I had trouble with the mathematics, but I was having lunch with a mathematician from the University of Tokyo named Takahasi and talked to him about it. He helped me out and together we solved the equations and published a paper. From it we could understand fluctuations in amplification by masers, and from that grew understanding of other kinds of amplifiers. So this finding of importance to engineering and physics came from biology and from my running into the right person.

I must also add some experiences which illustrate the problem with our getting fixed in a particular thought pattern. I myself was too fixed in my thought path about thermodynamics. I was even proud about knowing enough thermodynamics that I could show that molecules couldn't do the things I wanted to do. But in fact I was all wrong because I was too restricted in thinking that thermodynamics must apply. But still more striking is that there were many very prominent physicists who felt that I was just wrong about the maser being able to give a very pure frequency. While we were working on the first maser, Professor Llewellyn Thomas, famous for the Thomas Effect, kept telling me I was all wrong. I tried to talk with him about it, but he wouldn't bother to talk to me because he felt I didn't understand it. According to Professor Thomas, it couldn't give a pure frequency at all. After the maser was working, I was in Denmark and walking along the street with Neils Bohr. I am sure everybody knows who Neils Bohr was – the most famous physicist of quantum mechanics. He asked me what I was doing. That's what we all do with fellow scientists – "what are you doing"? We learn from that. I told him well, we had just built the maser, an oscillator of very pure frequency powered by molecules. Bohr said "no, no, that can't be right. There must be some mistake. That's not possible". I said we have actually made it and tried to explain why it gave a pure frequency. I am not sure he really understood me and believe the problem was that he was thinking of the uncertainty principle – something about which he was very knowledgeable. He recognized that the molecules go through a cavity in a finite time, which from a simple interpretation of the uncertainty principle means they can't have a very pure frequency. But he wasn't allowing for a large collection of molecules. Any engineer, if you told him you had a feedback oscillator, would immediately say "oh yes, that should give a rather pure frequency". But for a physicist, quantum mechanics and the uncertainly principle would say you can't get a pure frequency out of that. There has to be some uncertainty. John von Neumann, another famous physicist and mathematician whom I ran into at a cocktail party at Princeton University also asked me what I was doing. I told him that I had built this oscillator of very pure frequency. 'Oh no, that can't be right. You can't get such a pure frequency". I replied that we have it. "No, there is something wrong", von Neumann said; "you don't understand somehow". He went off to get another cocktail, but 15 minutes later came back and said "Hey, you are right". You see how our thoughts are too channelled. I believe all these great physicists had concentrated so much on the uncertainty principle of quantum mechanics, a very important principle, that anything even hinting at something different could not be right.

We all want to develop fields and find new things, but must recognize that many of these new things will entail surprises. We must explore. We must explore openly, we must take chances and recognize that we can't foresee everything. That's very difficult for an administrator or for a politician who must decide where to invest research money and effort in order to do what is useful, because new things cannot be very well foreseen. I once had a meeting with the chairman of our Congressional Committee on Science and Technology. We sat down for lunch with some other scientists. He wanted to talk to us about research support. This is the people's money he said, and we have to spend it on things which are clearly going to pay off because we have that responsibility. I responded to him and explained that important new things are not always predictable. I gave him an illustration of the laser's development, and after lunch he came to me and said "Should we have a special fund for crackpot ideas?" I said no, it's not crackpot ideas we need. What is needed is support for smart young people who are exploring new things, not crackpots. Other people may not agree with a creative person, but he or she is doing something that is interesting and is intelligent, that is the kind of case we need to support. We need flexibility. We also need interactions between people and interaction between fields. I frankly am somewhat concerned there is not as much interaction as there ought to be between our universities and our industry. We need to encourage interactions back and forth between industry and universities, between engineering and science, and between the different fields of science. Consider, for example, biology. Biology is becoming more fundamental in its research, calling more and more on physics and chemistry. And I am confident that as a result, it will turn up many interesting and important discoveries. We must all recognize the way science and technology really develop and allow for that in our policies and plans. They are a community development. A very large number of people have contributed to the development of lasers, and that's what has made the field grow rapidly, a growth that should continue. Openness to new ideas, explorations, and taking some chances are important. And the interactions of people with different ideas are important. We can hope that our societies recognize these aspects of the growth of science and technology and thus encourage their growth. The new surprises that will occur, as well as the systematic developments which are needed, will in the long run be very important to human welfare.

The Link Between Neutrino Masses and Proton Decay in Supersymmetric Unification

Jogesh C. Pati

University of Maryland at College Park, U.S.A

Fig. 1. Jogesh C. Pati delivering the B.M. Birla Science Centre Distinguished Lecture

Jogesh C. Pati was born in the Orissa State of India. After graduation with Honors from the famous Ravenshaw College of the Utkal University in 1955, Jogesh completed his Masters from the Delhi University in 1957. After obtaining his PhD from the University of Maryland in USA, he had a few very prestigious fellowships including the R.C. Tolman Post Doctoral position at Caltech from 1960 to 1962. Thereupon he joined as an Assistant Professor at the University of Maryland in 1963, becoming an Associate Professor in 1967 and a full Professor in 1973 at the same University. He was also the Chairman for the Centre for Theoretical Physics, University of Maryland between 1984 to 1987 and again 1993 to 1994.

Prof. Pati who is undoubtedly one of the foremost theoretical physicists of Indian origin has made pioneering contributions towards the goal of a unification of quarks and leptons, the fundamental particles, and of their gauge forces viz., strong interactions. His formulation done in collaboration

B.G. Sidharth (ed.), *A Century of Ideas*, 139–174.
© Springer Science+Business Media B.V. 2008

with Prof. Abdus Salam, in original gauge theory of quark lepton unification and their resulting insight that violations of baryon and lepton numbers, particularly which would manifest in proton decay are likely to be consequences of such a unification, has been a cornerstone of modern particle physics. The Pati-Salam $SU(4)$-color, left right symmetry and the associated existence of the righthanded neutrinos provides some of the ingredients for understanding the recently discovered neutrino oscillations and masses. Much of this work was done in the 1970s at the International Center for Theoretical Physics.

Prof. Pati has been the recipient of several honors and awards and has also held several prestigious Visiting Professorships. He was a Member of the Institute of Advanced Study, Princeton, a Visiting Professor or Scientist at ICTP, Trieste, CERN, Geneva, SLAC at Stanford, University of Bonn, the Schrodinger Visiting Professor of the University of Vienna, the B.M. Birla Visiting Professor at the B.M. Birla Science Centre and so on. He has also been a Guggenheim Fellow, the distinguished Homi J. Bhabha Chaired Professor of the Government of India and received the prestigious Dirac Medal in 2000. In honor of his life time contributions to theoretical elementary particle physics the University of Maryland organized a Special Symposium. He has well over a hundred important publications.

My acquaintance with Prof. Pati goes back to nearly thirty years. Through this period we have been meeting off and on. He is a very soft spoken and thorough analytical scholar whose words are always well measured. Another of his striking characteristics is his utter simplicity.

Following recent joint works with K. Babu and F. Wilczek, I stress here that supersymmetric unification, based on symmetries like $SO(10)$ or a string-derived $G(224) = SU(2)_L \times SU(2)_R \times SU(4)^C$ possesses some crucial features that are intimately linked to each other. They are: (a) gauge-coupling unification, (b) the masses and mixings of all fermions, including especially the neutrons, and last but not least (c) proton decay. In this context, it is noted that the value of $m(\nu_L) \sim 1/20eV$. suggested by the SuperK result, goes extremely well with the unification hypothesis, based on the ideas of (i) $SU(4)$ color, (ii) left-right symmetry and (iii) supersymmetry. A concrete proposal is presented within an economical $SO(10)$ framework that makes five successful predictions for the masses and mixings of the quarks and the charged leptons. The same framework explains why the $\nu_\mu - \nu_\tau$ oscillation angle is so large $(sin^2 2\Theta^{osc}_{\nu_\mu \nu_\tau} \approx 0.82 - 0.96)$ and yet V_{bc} is so small (≈ 0.04), both in accord with observation. The influence of the masses of the neutrinos and of the charged fermions on proton decay is discussed concretely, within the framework. The $\bar{\nu}K^+$ mode is expected to be dominant for SUSY $SO(10)$ as well as $SU(5)$. A distinctive feature of the $SO(10)$ model, however, is the likely prominence of the $\mu^+ K^\circ$ mode, which, for $SU(5)$, is highly suppressed. Our study shows that while current limits on the rate of proton decaying into $\bar{\nu}K^+$

---

[1] This is a technical talk.

is compatible with theoretical expectations, improvements in these limits by a factor of 5–10 should either turn up events, or else the $SO(10)$ framework described here, which is otherwise so successful, will be in jeopardy. Prominence of the $\mu^+ K^\circ$ mode, if observed, will be most significant in that it will reveal the intriguing link that exists between neutrino masses and proton decay in the context of supersymmetric unification.

1 Introduction

The SuperKamiokande (SK) result, convincingly showing the oscillation of ν_μ to ν_τ (or ν_X) with a value of $\delta m^2 \approx 10^{-2} - 10^{-3} eV^2$ and an almost maximal oscillation angle [1] $sin^2 2\Theta > 0.83$, clearly seems to require new physics beyond that of the standard model [2,3]. This, as well as the other relatively firm result of solar neutrino-deficit [4] serve as important clues to physics at a deeper level. Understanding these neutrino anomalies as well as the bizarre pattern of masses and mixings of the quarks and the charged leptons is a major challenge that ought to be met within a fundamental unified theory.

It is of course known that the ideas of grand unification [5–8], as well as those of superstrings [9] call for gauge coupling unification at a high scale and for nucleon-instability. Furthermore, both these features are known to acquire a new perspective [10, 11] in the context of supersymmetry [12]. (For recent reviews on this topic and relevant references see e.g. [13] and [14]). While proton decay is yet to show, the clearest empirical support in favor of grand unification and supersymmetry has so far come from the dramatic meeting of the three gauge couplings of the standard model that is found to occur at a scale of $M_X \approx 2 \times 10^{16} GeV$, when these couplings are extrapolated from their measured values at LEP to high energies, in the context of supersymmetry [10].

One major goal of this talk will be to stress that supersymmetric unification based on symmetries like $SO(10)$ [15], or (for most purposes) a string-derived [6, 16] $G(224) = SU(2)_L \times SU(2)_R \times SU(4)^C$, has implications not only for (i) gauge coupling unification and (ii) proton decay, but also for (iii) the masses and mixings of the charged fermions, as well as for (iv) those of the neutrinos. In fact, within a unified theory, all four features (i)–(iv) get intimately linked to each other, much more so than commonly thought. Each of these, including even charged fermion and neutrino-masses, provides some essential clue to the nature of higher unification. As regards the link between the four features, even neutrino masses turn out to have direct influence on proton decay. This is because the latter receives important contributions through a new set of $d = 5$ operators that depend directly on the Majorana masses of the right-handed neutrinos [17]. These new $d = 5$ operators, which were missed in the literature, contribute significantly to proton decay amplitudes, in addition of course to the "standard" $d = 5$ operators [11], which arise through the exchange of the color-triplet Higgsinos related to the electro-weak doublets. The standard and the new $d = 5$ operators, related to the charged fermion as well as the neutrino masses, together raise our expectation that proton decay should be observed in the near future [18].

I elucidate these remarks in the next four sections, covering the following topics:

(1) I first recall briefly the motivations for left-right symmetric unified theories, utilizing neutrino masses suggested by the SuperKamiokande result, as a guide. The support for supersymmetric unification in the light of the LEP data is noted. Further, the origin of such a unification in the context of superstrings as well as the potential problem of rapid proton decay that arises within supersymmetric theories are briefly reviewed. These discussions provide the background needed to cover the materials in the remaining sections.

(2) I then present arguments [2] to show that the SuperK result, especially the observed δm^2, interpreted as $m(\nu_\tau)^2$, receives a simple and natural explanation within the ideas of higher unification based on the symmetry group $G(224)$ [6], and thus $SO(10)$ or E_6. Such an explanation would not be possible within $SU(5)$.

(3) I present the first part of a recent work by Babu, Wilczek and myself [18], in which we attempt to understand, in the context of supersymmetric $SO(10)$, the masses and mixings of the neutrinos, suggested by the atmospheric and the solar neutrino anomalies, in conjunction with those of the quarks and the charged leptons. Adopting familiar ideas of generating hierarchical eigenvalues through off-diagonal mixings, and correspondingly cabibo-like mixing angles we find that the bizarre pattern of masses and mixings observed in the charged fermion sector, remarkably enough, can be adequately described (with $\sim 10\%$ accuracy) within an economical and thus predictive $SO(10)$ framework. A concrete proposal is presented involving a minimal Higgs system that provides five successful predictions for the masses and mixings of the quarks and the charged leptons in the three families. The same description goes extremely well with a value of $m(\nu_\tau) \sim (1/20)eV$ as well as with a large $\nu_\mu - \nu_\tau$ oscillation angle $(sin^2 2\Theta_{\nu_\mu \nu_\tau}^{osc} \approx 0.82 - 0.96)$, despite highly non-degenerate masses of the light neutrinos. Both these features are in good agreement with the SuperK result. Furthermore, this framework generically seems to support the small angle MSW explanation for the solar neutrino deficit [19].

I next present the second part of the work by Babu, Wilczek and myself [18] in which we link the rather successful supersymmetric $SO(10)$ framework describing fermion masses (noted above), with expectations for proton decay. We find that, given the SuperK result that suggests $m(\nu_\tau) \sim (1/20)eV$ and a large oscillation angle, the contribution from the new $d = 5$ operators mentioned above, and to some extent that from the standard operators as well, are significantly enhanced. As a result, in spite of generous allowance for uncertainties in the matrix elements and the SUSY spectrum, the inverse decay rate for the dominant $\bar{\nu}K^+$ mode is found to be bounded from above by about 7×10^{33} years. Typically, the lifetime should of course be lower than this bound. Furthermore, the $\mu^+ K^\circ$ mode is found to be prominent, with a branching ratio typically in the range of 10–50%, entirely because of contribution from the new operators. For comparison, minimal SUSY $SU(5)$, which has only the standard operators, typically leads to branching ratios $\leq 10^{-3}$ for this mode. Thus, our study of proton decay, correlated with fermion masses, strongly suggest that at least the candidate events for proton decay should be observed in the very near future, already at SuperK. The $\mu^+ K^\circ$ mode, if observed, would be specially important in exhibiting the link between neutrino masses and proton decay that exists within the $G(224)/SO(10)$ route to supersymmetric unification [18].

2 Learning from Neutrino Masses About Higher Unification

2.1 Motivations for $SU(4)$ Color Left-Right Symmetric Theories

If one assumes a hierarchical pattern of masses for the light neutrinos (with $m_{\nu_e} \ll m_{\nu_\mu} \ll m_{\nu_\tau}$), which goes well within a quark-lepton unified theory, the SuperK result interpreted as $\nu_\mu - \nu_\tau$ oscillation, suggests a value for the ν_τ mass: $m_{\nu_\tau} \approx 1/20 eV ((1/2)$ to 2$)$. One can argue, as shown later in this section (see also [2]), that a ν_τ mass of this order can be understood simply within supersymmetric unified theories which are forced to introduce the existence of right-handed (RH) neutrino, accompanying the observed left-handed ones. Postponing an estimate of the ν_τ mass for a moment, if one asks the question: What symmetry on the one hand dictates the existence of the RH neutrinos, and on the other hand also ensures quantization of electric charge, together with quark-lepton unification, one is led to two very beautiful conclusions:

(i) Quarks and leptons must be unified minimally within the symmetry $SU(4)$ color, and that,
(ii) deep down, the fundamental theory should possess a left-right symmetric gauge structure: $SU(2)_L \times SU(2)_R$.

In short, the standard model symmetry must be extended minimally to the gauge symmetry [5,6],

$$G(224) = SU(2)_L \times SU(2)_R \times SU(4)^C \tag{1}$$

With respect to $G(224)$, all members of the electron family fall into the neat pattern:

$$F^c_{LR} = \begin{bmatrix} u_x & u_y & u_b & \nu_v \\ d_x & d_y & d_b & c^- \end{bmatrix}_{LR} \tag{2}$$

The left-right conjugate multiplets F^c_L and F^c_R transform as (2,1,4) and (1,2,4) respectively, with respect to $G(224)$; likewise for the mu and the tau families.

Viewed against the background of the standard model, the symmetry structure $G(224)$ brought some attractive features to particle physics which include:

(i) Organization of all members of a family $(8_L + 8_R)$ within one left-right self-conjugate multiplet, with their peculiar hypercharges fully explained.

(ii) Quantization of electric charge, explaining why $Q_{\text{electron}} = -Q_{\text{proton}}$

(iii) Quark-lepton unification through $SU(4)$ color.

(iv) Left-right (i.e. parity) and particle-antiparticle symmetries in the fundamental laws which are violated only spontaneously [6, 20]. Thus, within the symmetry structure $G(224)$, quark-lepton distinction and parity violation may be viewed as low energy phenomena which should disappear at sufficiently high energies.

(v) Existence of right-handed neutrinos: Within $G(224)$, there must exist a right-handed (RH) neutrino (ν_R) accompanying the left-handed one (ν_L) for each family because ν_R is the fourth color partner of the corresponding RH up-quarks. It is also the $SU(2)_R$-doublet partner of the associated RH charged lepton (see eq. (2)). The RH neutrinos seem to be essential now (see later discussions) for understanding the non-vanishing light masses of the neutrinos, as suggested by the recent observations of neutrino oscillations.

(vi) B-L as a local gauge symmetry: $SU(4)$ color introduces B-L as a local gauge symmetry. Thus following the limits from Eotvos experiments, one can argue that B-L must be violated spontaneously. It has been realized, in the light of recent works, that to implement baryogenesis in spite of electro-weak sphaleron effects, such spontaneous violation of B-L at high temperatures may well be needed [21].

2.2 Route to Higher Unification: $SU(5)$ versus $G(224)/SO(10)$

To realize the idea of a single gauge coupling governing the three forces [5,6], one must embed the standard model symmetry or $G(224)$, into a simple (or effectively simple, like $SU(N) \times SU(N)$) gauge group. The smallest such group is $SU(5)$ [7] which contains the standard model symmetry but not $G(224)$. As a result, $SU(5)$ does not possess some of the main advantages of $G(224)$ listed above. In particular, $SU(5)$ splits members of a family into two multiplets: $5 + 10$, whereas $G(224)$, subject to L-R symmetry, groups them into just one multiplet. $SU(5)$ violates parity explicitly. It does not possess $SU(4)$ color and therefore does not gauge B-L as a local symmetry. Further, $SU(5)$ does not contain the RH neutrinos as an integral feature. As I will discuss below, these distinctions between $SU(5)$ versus $G(224)$, or its extensions (see below), turn out to be especially relevant to considerations of neutrino as well as charged fermion masses, and thereby to those of proton decay.

Since $G(224)$ is isomorphic to $SO(4) \times SO(6)$, the smallest simple group to which it can be embedded is $SO(10)$ [15]. Historically, by the time $SO(10)$ was proposed, all the advantages of $G(224)$ [(i)–(vi), listed above] and the ideas of higher unification were in place. Since $SO(10)$ contains $G(224)$, the features (i)–(vi) are of course retained by $SO(10)$. In addition, the 16-fold left-right conjugate set $(F_L^c + F_R^c)$ of $G(224)$ corresponds to the spinorial 16 of $SO(10)$. Thus, $SO(10)$ preserves even the 16-plet family-structure of $G(224)$, without a need for any extension. If one extends $G(224)$ to the still higher symmetry [6] E_6, the advantages (i)–(vi) are retained, as in $SO(10)$, but in this case, one must extend the family structure from a 16 to a 27-plet.

Comparing $G(224)$ with $SO(10)$ as mentioned above, $SO(10)$ possesses all features (i)–(vi) of $G(224)$; in addition it offers gauge coupling unification. I should, however, mention at this point that the perspective on coupling unification and proton decay has changed considerably in the context of supersymmetry and superstrings. In balance, a string-derived $G(224)$ offers some advantages over a string-derived $SO(10)$, while the reverse is true as well. Thus, it seems that a definite choice of one over the other, as an effective theory below the string scale, is hard to make at this point. I will return to this point shortly.

2.3 Gauge Coupling Unification: Need for Supersymmetry

It has been known for some time that the precision measurements of the standard model coupling constants (in particular $sin^2\Theta_W$) at LEP put severe constraints on the idea of grand unification. Owing to these constraints, the non-supersymmetric minimal $SU(5)$, and for similar reasons, the one step breaking minimal non-supersymmetric $SO(10)$ model as well, are now excluded [23].

But the situation changes radically if one assumes that the standard model is replaced by the minimal supersymmetric standard model (MSSM), above a threshold of about $1TeV$. In this case, the three gauge couplings are found to meet [10], at least approximately, provided $\alpha_3(m_z)$ is not too low (see figures in [13,23]). Their scale of meeting is given by

$$M_X \approx 2 \times 10^{16} GeV \quad (\text{MSSM} \quad \text{or} \quad \text{SUSYSU}(5)) \tag{3}$$

M_X may be interpreted as the scale where a supersymmetric grand unification symmetry (GUT) (like minimal SUSY $SU(5)$ or $SO(10)$) – breaks spontaneously into the supersymmetric standard model symmetry $SU(2)_L \times U(1) \times SU(3)^C$.

The dramatic meeting of the three gauge couplings thus provides a strong support for both grand unification and supersymmetry.

2.4 Compatibility Between MSSM and String-Unifications

The superstring theory [9], and now the M theory [24] provide the only known framework that seems capable of providing a good quantum theory of gravity as well as a unity of all forces, including gravity. It thus becomes imperative that the meeting of the gauge couplings of the three non-gravitational forces which occur by the extrapolation of the LEP data in the context of MSSM, be compatible with string unification.

Now, string theory does provide gauge coupling unification for the effective gauge symmetry, below the compactification scale. The new feature is that even if the effective symmetry is not simple, like $SU(5)$ or $SO(10)$, but instead is of the form $G(213)$ or $G(224)$ (say), the gauge couplings of $G(213)$ or $G(224)$ should still exhibit familiar unification at the string-scale, for compactification involving appropriate Kac-Moody levels (i.e. $k_2 = k_3 = 1, k_Y = \frac{5}{3}$ for $G(213)$), barring of course string threshold corrections [25]). And even more, the gauge couplings unify with the gravitational coupling $(8\pi G_N/\alpha')$ at the string scale, where G_N is the Newton's constant and α' is the Regge slope.

Thus one can realize coupling unification without having a GUT-like symmetry below the compactification scale. This is the new perspective brought forth by string theory. There is, however, an issue to be resolved. Whereas the MSSM unification scale, obtained by extrapolation of low energy data is given by $M_N \approx 2 \times 10^{16} GeV$, the expected one-loop level string unification scale [25] of $M_{st} \approx g_{st} \times (5.2 \times 10^{17} GeV) \approx 3.6 \times 10^{17} GeV$ is about twenty times higher. Here, one has used $\alpha_{st} \approx \alpha_G UT(MSSM) \approx 0.04$.

Possible resolutions of this mismatch between M_N and M_{st} by about a factor of 20 have been proposed (for a comprehensive review see e.g. [13] and [14]). These include:

(i) utilizing the idea of string duality that allows a lowering of M_{st} [26] compared to the value suggested by [25]; alternatively
(ii) the idea of a semi-perturbative unification that assumes the existence of two vector-like families at the TeV scale, $(16+16)$ which raise α_{GUT} to about $0.25 - 0.3$, and thereby also M_X to a few $\times 10^{17} GeV$ [27]; or
(iii) the alternative of a string GUT solution, which would arise if superstrings yield an intact grand unification symmetry like $SU(5)$ or $SO(10)$, together with supersymmetry and the right spectrum – i.e. three chiral families and a suitable Higgs system – at M_{st}, and if the symmetry would break spontaneously at $M_X \sim 1/20 M_{st}$ to the standard model symmetry. In this last case, the gauge couplings would run together between M_X and M_{st} and thus the question of a mismatch between the two scales would not even arise. However, as yet, there does not seem to be even a semi-realistic string-derived GUT model [28]. Further, to-date, no string GUT solution exists with a resolution of the well-known doublet splitting problem, without which one would face the problem of rapid proton decay through the $d = 5$ operators [11] (see discussions below). This does not necessarily mean that a realistic GUT solution exhibiting doublet-triplet splitting cannot ultimately emerge from the string or the M theory.

While each of the solutions mentioned above possesses a certain degree of plausibility (see [13] for some additional possibilities), it is not clear, which, if any is utilized by the true string vacuum. This is related to the fact that, as yet, there is unfortunately no insight as to how the true vacuum is selected in the string or in the M theory.

2.5 A GUT or a Non-GUT String Solution?

Comparing string-derived GUT solutions with non-GUT solutions, where the former yield symmetries like $SU(5)$ or $(SO(10)$, while the latter lead to symmetries like $G(213)$ or $G(224)$ at the string scale, we see from the discussions above that each class has a certain advantage and possible disadvantages as well, compared to the other. In particular, a string GUT solution has the positive feature, explained above, that the issue of a mismatch between M_{st} and M_X does not arise for such a solution. For a non-GUT solution, however, although plausible mechanisms of the type mentioned above could remove the mismatch, a priori it is not clear whether any such mechanism is realized.

On the other hand, for a string-derived GUT solution [28], achieving doublet-triplet splitting so as to avoid rapid proton decay, is still a major burden. In this regard, the non-GUT solutions possess a distinct advantage because the dangerous color triplets are often naturally projected out [29,30]. Furthermore, these solutions invariably possess new "flavor" gauge symmetries, which are not available in GUTs. The flavor symmetries turn out to be immensely helpful in (a) providing the desired protection against gravity induced rapid proton decay [31], (b) resolving certain naturalness problems of supersymmetry such as those pertaining to the issues of squark-degeneracy, neutrino-Higgsino mixing and CP violation [32]–[34], and (c) explaining qualitatively the observed fermion mass hierarchy [29].

Weighing the advantages and possible disadvantages of both, it seems hard at present to make a clear choice between a GUT versus a non-GUT string solution. We will therefore keep our options open and look for other means, for example certain features of proton decay and neutrino masses, to provide a distinction. We will thus proceed by assuming that for a GUT solution, string theory will somehow provide a resolution of the problem of the doublet-triplet splitting, while for a non-GUT string solution, we will assume that one of the mechanisms mentioned above (for instance, that based on string-duality [26]), does materialize removing the mismatch between M_X and M_{st}. In general, a combination of the two mechanisms [26, 27] may also play a role.

It turns out that there are many similarities between the predictions of $SO(10)$ and of a string-derived $G(224)$, especially as regards neutrino and charged fermion masses, primarily because both contain $SU(4)$ color.

With these discussions on higher unification, including the ideas of supersymmetry and superstrings to serve as a background, I proceed to discuss more concretely, firstly the masses and mixings of all fermions, and finally, their link to proton decay. An estimate of m_{ν_τ}, is presented next.

3 Mass of ν_τ: An Evidence in Favor of the $G(224)$ Route

One can now obtain an estimate for the mass ν_L^τ in the context of $G(224)$ or $SO(10)$ by using the following three steps [2]:

(i) First, assume that B-L and I_3R, contained in a string-derived $G(224)$ or $SO(10)$, break near the unification scale:

$$M_X \sim 2 \times 10^{16} GeV, \tag{4}$$

through VEVs of Higgs multiplets of the type suggested by string solutions [35]– i.e. $< (1,2,4)_H >$ for $G(224)$ or $< 16_H >$ for $SO(10)$, as opposed to 126_H. In the process, the RH neutrinos (ν_R^i), which are singlets of the standard model, can and generically will acquire superheavy Majorana masses of the type $M_R^{ij} \nu_R^{iT} C^{-1} \nu_R^j$, by utilizing the VEV of $< 16_H >$ and effective couplings of the form:

$$L_M(SO(10)) = \int_R^{ij} 16_i, 16_j 16_H, 16_H / M + hc \tag{5}$$

A similar expression holds for $G(224)$. Here $i, j = 1, 2, 3$, correspond respectively to e, μ and τ families. Such gauge-invariant non-renormalizable couplings might be expected to be induced by Planck-scale physics involving quantum gravity or string effects and/or tree-level exchange of superheavy states, such as those in the string tower. With f_{ij} (at least the largest among them) being of order unity, we would thus expect M to lie between $M_{\text{Planck}} \approx 2 \times 10^{18} GeV$ and $M_{\text{string}} \approx 4 \times 10^7 GeV$. Ignoring for the present off-diagonal mixing (for simplicity), one thus obtains:

$$M_{3R} \approx \frac{f_{33} < 16_H >^2}{M} \approx f_{33}(2 \times 10^{14} GeV)\eta^2(M_{\text{Planck}}/M) \tag{6}$$

This is the Majorana mass of the RH tau neutrino. Guided by the value of M_X, we have substituted $<16_H>= (2 \times 10^{16} GeV)\eta$ where $\eta \approx 1/2$ to 2, for this estimate.

(ii) Second, assume that the effective gauge symmetry below the string scale contains $SU(4)$ color. Now using $SU(4)$ color and the Higgs multiplet $(2,2,1)_H$ of $G(224)$ or equivalently 10_H of $SO(10)$, one obtains the relation $m_\tau(M_X) = m_b(M_X)$, which is known to be successful. Thus, there is a good reason to believe that the third family gets its masses primarily from the 10_H or equivalently $(2,2,1)_H$. In turn, this implies:

$$m(\nu_{\text{Dirac}}^\tau) \approx m_{\text{top}}(M_X) \approx (100 - 120)GeV \tag{7}$$

Note that this relationship between the Dirac mass of the tau neutrino and the top mass is special to $SU(4)$ color. It does not emerge in $SU(5)$.

(iii) given the superheavy Majorana masses of the RH neutrinos as well as the Dirac masses as above, the see-saw mechanism [36] yields naturally light masses for the LH neutrinos. For ν_L^τ (irgnoring mixing), one thus obtains, using eqs. (6) and (7),

$$m(\nu_L^\tau) \approx \frac{m(\nu_{\text{Dirac}}^\tau)}{M_{3R}} \approx [(1/20)eV(1 \text{ to } 1.44)/f_{33}\eta^2](M/M_{\text{Planck}}) \tag{8}$$

Considering that on the basis of the see-saw mechanism, we naturally expect that $m(\nu_L^e) \ll m(\nu_L^\mu) \ll m(\nu_L^\tau)$, and assuming that the SuperK observation represents $\nu_L^\mu - \nu_L^\tau$ (rather than $\nu_L^\mu - \nu_x$) oscillation, so that the observed $\delta m^2 \approx 1/2(10^{-2} - 10^{-3})eV^2$ corresponds to $m(\nu_L^\tau)_{obs} \approx (1/15 \text{ to } 1/40)eV$, it seems truly remarkable that the expected magnitude of $m(\nu_L^\tau)$, given by eq. (8), is just about what is observed if $f_{33}\eta^2(M_{\text{Planck}}/M)$ seems most plausible and natural [2]. It should be stressed that the estimate (8) utilizes the ideas of both supersymmetric unification, which yields the scale of M_{3R} (eq. (6)), and of $SU(4)$ color that yields $m(\nu_{\text{Dirac}}^\tau)$ (eq. (7)). The agreement between the expected and the SuperK result thus suggests that, at a deeper level, near the string or the coupling unification scale M_X, the symmetry group $G(224)$ and thus the ideas of $SU(4)$ color and left-right symmetry are likely to be relevant to nature.

By providing clear support for $G(224)$, the Super K result selects out $SO(10)$ or E_6 as the underlying grand unification symmetry, rather than $SU(5)$. Either $SO(10)$ or E_6 or both of these symmetries ought to be relevant at some scale, and in the string context, as discussed in Section 2, that may well be in higher dimensions, above the compactification scale, below which there need be no more than just the $G(224)$ symmetry. If, on the other hand, $SU(5)$ were regarded as a fundamental symmetry, first, there would be no compelling reason, based on symmetry alone, to introduce a ν_a because it is a singlet of $SU(5)$. Second, even if one did introduce ν_R^i by hand, their Dirac masses, arising from the coupling $h^i 5_i < 5_H > \nu_R^i$, would be unrelated to the up-flavor masses and thus rather arbitrary (contrast with eq. (7)). So also would be the Majorana masses of the $\nu_R^i S$, which are $SU(5)$ invariant and thus can even be of order Planck scale (contrast with eq. (6)). This would give $m(\nu_L^\tau)$ in gross conflict with the observed value. In this sense, the SuperK result appears to disfavour $SU(5)$ as a fundamental symmetry, with or without sypersymmetry.

4 Fermion Masses and Neutrino Oscillations in $SO(10)$

4.1 Preliminaries

I now discuss the masses and mixing of the quarks and charged leptons in conjunction with those of the neutrinos, to see first of all how well they can be understood together within the ideas of higher unification.

The most striking regularity in the masses of the fermions belonging to the three families (at least of the charged ones) is their inter-family hierarchy. This is reflected by the uniform pattern: $m_t \gg m_c \gg m_u; m_b \gg m_s \gg m_d;$ and $m_\tau \gg m_\mu \gg m_e$. Apart from this gross feature however, if one examines the pattern in more detail, it looks rather bizarre, especially when one compares intra-family mass splittings of the three families. For instance, while $m_t^\circ/m_b^\circ \sim 60$, one finds that $m_c^\circ/m_s^\circ \sim 10$ and $m_u^\circ/m_d^\circ \sim 1/2$. Here, the superscript $\circ$ denotes that the respective mass is evaluated at the unification scale. Note that the ratio of the up - and down-flavor masses within a family varies widely in going from the third to the second to the first family. Further, comparing quark versus lepton masses of the down-flavor within a family in contrast to $m_b^\circ \approx m_\tau^\circ$, that suggests $b - \tau$ unification for the third family, one finds: $m_s^\circ \sim m_\mu^\circ/3$ and $m_d^\circ \sim 3m_e^\circ$ [37]. In short, there does not seem to be any obvious regularity in the intra-family mass splittings. The question is: do these apparent irregularities still have a simple origin?

The pattern seems to be equally bizarre when one examines the mixing angles. While the parameter $V_{us} = \Theta_c$, representing the mixing between the electron and the muon families in the quark sector, is moderately large (≈ 0.21), the parameter V_{cb}, representing $\mu - \tau$ family mixing, also in the quark sector, is small (≈ 0.044). This feature seems even more strange, when one compares V_{cb} with the $\nu_\mu - \nu_\tau$ oscillation angle, which also represents $\mu - \tau$ family mixing, although in the leptonic sector. This angle seems to be almost maximal: $sin^2 2\Theta_{\nu_\mu \nu_\tau}^{osc} > 0.83$. One might have been tempted to associate such a large mixing angle with near degeneracy of ν_μ and ν_τ, as has been attempted by several authors. But, then, such degeneracy does not go well with the see-saw formula, especially within a unified scheme in which the Dirac masses of the neutrinos are related to those of the quarks which exhibit a large inter-family hierarchy. Thus one major puzzle is: Why V_{bc} is so small and yet $\Theta_{\nu_\mu \nu_t}^{osc}$ so large? Could the smallness of one imply the largeness of the other within a quark-lepton unified theory? Further, are these peculiarities of the mixing angles related to the irregularities in the intra-family mass splittings mentioned above?

From a theoretical viewpoint, the goal is to resolve some of these puzzles within a unified predictive theory, in particular to understand the masses and mixing of the neutrinos in conjunction with those of the quarks and the charged leptons, rather than in isolation. It is however known that there is no obvious way to address any of these puzzles in the context of the standard model (SM), because, a priori, the SM allows for all the masses and mixings to be arbitrary parameters. Even ignoring CP violation for the present discussion, there are 12 such observables: $m_t, m_b, m_\tau, m_c, m_s, m_\mu, m_u, m_d, m_l,$ V_{us}, V_{cb} and V_{ub}. The 3×3 mass matrices of the 3 sectors (up, down and charged lepton) would in general have as many as $9 \times 3 = 27$ real parameters, which represent, however, only 12 observables. The parameters would even increase if one introduces RH neutrinos and considers both the Dirac and the Majorana mass matrices of the three neutrinos.

To reduce the number of parameters, it thus seems that one may have to appeal to symmetries of two kinds: first like those in $G(224)$ or $SO(10)$, which relate quark versus lepton as well as up-versus down-Yukawa couplings, and second "flavor" symmetries which distinguish between the three families (c, μ and τ) and could account for inter-family mass hierarchy. Interestingly enough, these latter symmetries do seem to arise in string solutions [29, 30] though not in GUT's.

To proceed further, we will use the following guidelines.

(1) Hierarchy through off-diagonal mixings: Recall earlier attempts [38] that attribute hierarchy in the quark mass matrices of the first two families to matrices of the type:

$$M = \begin{pmatrix} 0 & \epsilon \\ \epsilon & l \end{pmatrix} m_s^{(0)},$$ (9)

for the (d, s) quarks, and likewise for the (u, c) quarks. Here $\epsilon \sim 1/10$. Note the symmetric form of eq. (9) (i.e. $M_{12} = M_{21}$) and especially the hierarchical pattern: $(1, 1) \ll (1, 2) \ll (2, 2)$, where $(1, 1) \leq 0(\epsilon^3)$. The symmetric nature of eq. (9) is guaranteed by group theory if the relevant Higgs field is a **10** of $SO(10)$. The hierarchical entries in eq. (9) can be ensured by imposing a suitable flavor symmetry that distinguishes between the two families (origin of such symmetries must ultimately be attributed to, for example, string theory). The pattern (eq. (9)) has the virtues that (a) it generates a hierarchy larger than the input parameter $\epsilon : |m_d/m_s| \approx \epsilon^2 \ll \epsilon$, and (b) it leads to the rather successful expression for the Cabibo angle:

$$\Theta_C \simeq \left| \sqrt{\frac{m_d}{m_s}} - e^{\imath\phi} \sqrt{\frac{m_u}{m_c}} \right|$$ (10)

Using $\sqrt{m_d/m_s} \simeq 0.22$ and $\sqrt{m_u/m_c} \simeq 0.06$, we see that eq. (7) works within 30% for any value of the phase ϕ, and perfectly for a value of the phase parameter ϕ around $\pi/2$.

A generalization of the pattern (eq. (9)) to the case of three families would suggest that the first and the second families (i.e. the e and the μ families) receive their masses primarily through their mixings with the third family (τ); the $(3, 3)$ – element in this case is then the leading one in each sector. One must also rely on flavor symmetries that distinguish between the e, μ and τ families so as to ensure that the $(1, 3)$ and $(1, 2)$ mixing elements are smaller than the $(2, 3)$ – element. We will follow this guideline, except, however, for the modification noted below.

(2) The need for an antisymmetric component: Although the symmetric hierarchical mass matrix (9) works well for the first two families, a matrix of the same form fails altogether to reproduce V_{cb}, for which it would yield:

$$|V_{cb}| \simeq \left| \sqrt{\frac{m_c}{m_t}} - e^{\imath\chi} \sqrt{\frac{m_s}{m_b}} \right|$$ (11)

Given that $\sqrt{m_s/m_b} \simeq 0.17$ and $\sqrt{m_c/m_t} \simeq 0.06$, we see that eq. (11) would yield $|V_{cb}|$ varying between 0.11 and 0.23, depending upon the value of the phase χ. This is however too big compared to the observed value of $V_{cb} \approx 0.04 \pm 0.003$, by at least a factor of 3. We thus see that the simple square root formula for the mixing angle in each sector ($sin\Theta_{ij} \approx tan\Theta_{ij} = \sqrt{m_i/m_j}$;

(see eq. (10) or (11)), arising from a symmetric matrix of the form eq. (9), fails for V_{cb}. We would interpret this failure as a clue to the presence of antisymmetric contribution to off-diagonal mixing in the mass matrix together with a symmetric one, (thus $m_{ij} \neq m_{ji}$) which would modify the square-root formula for the mixing angle to $\sqrt{(m_i/m_j)}\sqrt{(m_{ij}/m_{ji})}$, where m_i and m_j denote the respective eigenvalues. We will note below a simple group theoretical origin of such an antisymmetric component in $SO(10)$, even for a minimal Higgs system, and point out its crucial role in resolving some of the puzzles alluded to, above. The resolution would depend, however, on an additional feature noted below.

(3) The need for a contribution proportional to B-L: The success of the relations $m_b^\circ \approx m_t^\circ$ and also $m_\tau^\circ \approx m(\nu_\tau)^\circ_{\text{Dirac}}$ suggests that the members of the third family receive their masses primarily from the VEV of a Higgs field, which is a singlet of $SU(4)$ color and thus independent of B-L. That is in fact the case for the Higgs transforming as (2,2,1) of $G(224)$ or 10 of $SO(10)$. However, the empirical observations of $m_s^\circ \sim m_\mu^\circ/3$ and $m_d^\circ \sim 3m_e^\circ$, as well as the suppression of V_{bc} (noted above) together with the enhancement of $\Theta^{osc}_{\nu_\mu \nu_\tau}$, (SuperK result) clearly calls for a contribution proportional to B-L as well. This would be the case for contributions from the VEV of a Higgs transforming as 15 of $SU(4)$ color. We note below how such a contribution can arise simply for a minimal Higgs system in $SO(10)$. The amusing thing is that such a contribution, while it is proportional to B-L, turns out to be anti-symmetric as well, in the family-space, fulfilling the need (2).

I now present, following [18], a simple and predictive mass matrix, based on $SO(10)$, which is constructed by using the guidelines (1)–(3). For simplicity, I first consider only the μ and the τ families. The discussion is extended later to include the electron family.

4.2 The Minimal Higgs System for $SO(10)$ Breaking and Fermion Masses

The minimal Higgs system, capable of breaking $SO(10)$ at the unification scale M_X into the SM symmetry $G(213)$ consists of a 45_H, a 16_H and (for supersymmetry) a $\overline{16}_H$. Of these $<45_H> \sim M_X$ breaks $SO(10)$ into $G(2213) = SU(2)_L \times SU(2)_R \times U(1)_{B-l} \times SU(3)^C$, while $<16_H> = <\overline{16}_H> \sim M_X$ breaks $SU(2)_R$ and B-L and thus $G(2213)$ into $G(213)$. To break $G(213)$ into $U(1)_{em} \times Su(3)^C$ at the electro-weak scale, one minimally needs in addition the VEV of a 10_H. Thus the minimal Higgs system, that is needed for appropriate $SO(10)$ breaking, consists of the set:

$$H_{\text{minimal}} = \{45_H, 16_H, \overline{16}_H, 10_H\} \tag{12}$$

Of these, only 10_H can have Yukawa coupling with the fermions at the cubic level of the form $h_{ij}16_i 16_j 10_H$, which could be the dominant source of masses, especially for fermions belonging to the third family. But the first two families must have additional sources for their masses because $a < 10_H >$ by itself would lead to three undesirable results: (a) $V_{CKM} = 1$, (b) purely symmetric mass-matrices, and (c) (B-L)-independent masses. We have on the other hand argued above the antisymmetric and (B-L)-dependent contributions to mass matrices are needed.

Now, there exist large-dimensional tenrosial multilets of $SO(10)$, that is 126_H and 120_H, which can have cubic-level Yukawa couplings with the fermions and give (B-L)-dependent contributions. Further, $< 120_H >$ gives purely family-antisymmetric contributions, as needed. There are however, two a priori reasons why we prefer not to use these large-dimensional multiplets: (a) They seem to be hard, if not impossible, to emerge from string solutions [35], and (b) generically, such large-dimensional multiplets tend to give large threshold corrections (typically exceeding 20%) to $\alpha_3(m_Z)$, thereby rendering observed coupling unification fortuitous. By contrast, the multiplets in the minimal set can arise in string solutions leading to $SO(10)$ $((45_H)$ arises at Kac-Moody level ≥ 2, while $16_H, \bar{16}_H$ and 10_H arise at level 1), and their threshold corrections have been computed. They were found not only to be smaller in magnitude, but also to have the right sign to go well with observed coupling unification [18].

Given these advantages of the minimal Higgs system (compared to those containing large multiplets like 126_H and/or 120_H) for $SO(10)$ breaking, the question arises: can this minimal system meet the requirements arising from fermion masses and mixing – that is, (a) $V_{CKM} \neq 1$, (b) presence of antisymmetric, and (c) that of (B-L)-dependent contributions? It was noted in [18] that minimal Higgs system can indeed meet all three requirements quite simply, if one allows for not just cubic, but also (seemingly) non-renormalizable effective quartic couplings of this minimal set with the 16-plets of fermions. Such quartic couplings could well arise through exchanges of superheavy particles (for example those in the string-tower) involving renormalizable couplings, and/or through quantum gravity.

Allowing for such cubic and quartic couplings of the minimal Higgs system and adopting the guideline eq. (1) of family hierarchical couplings, we are led to suggest the following effective Lagrangian for generating masses and mixings of the μ and τ families [18]. (The same consideration is extended later to include the electron family. For a related but different pattern, see [39]).

$$L_{\text{Yukawa}} = h_{33} 16_3 16_3 10_H + \frac{a_{23}}{M} 16_2 16_3 10_H 45_H$$

$$+ \frac{g_{23}}{M} 16_2 16_3 16_H + h_{23} 16_2 16_3 10_H \tag{13}$$

Note that a mass matrix of the type shown in eq. (9) (barring its symmetric form) results if the first term $h_{33} < 10_H >$ is dominant. This ensures $m_b^\circ \approx m_\tau^\circ$ and $m_\tau^\circ \approx m(\nu_{\text{Dirac}}^\tau)^\circ$.

The smallness of the remaining terms responsible for all-diagonal mixings, by about an order of magnitude compared to the h_{33} term, may come about as follows. First, as mentioned before, the smallness of the $SO(10)$ invariant coupling $h_{22}16_210_H$ (not shown) compared to the h_{23} coupling and that of h_{23} compared to h_{33} (i.e. $h_{22} \ll h_{21}llh_{33}$) may well have its origin in a flavor symmetry (or symmetries), which assigns different charges to the three different families, and also to the Higgs-like fields. In this case, assuming that the h_{33} term is allowed by the flavor symmetries and that the second and the third families have different flavor charges, the h_{23} term will not be allowed as a genuine cubic coupling. It can still arise effectively by utilizing an effective non-renormalizable coupling $h_{23}16_216_310_H < S > /M$ where S is an $SO(10)$-singlet carrying appropriate flavor charge(s), and acquires a $VEV \sim M_U$. In this case, $h_{23}(= h_{23} < S > /M)$ can naturally be $O(1/10)h_{23}$, if $h_{23} \sim h_{33}$ and $< S > /M \sim M_U/M_{st} \sim 1/10$. The h_{22} term would then be suppressed by $(< S > /M)^2 \sim 10^{-2}$, compared to h_{23}, as desired. Now, as regards the effective non-renormalizable terms in eq. (13), assuming that they are generated by quantum gravity or stringy effects and/or by tree-level exchanges of superheavy states (see e.g. those in the string tower), the scale M is naturally expected to be of order $M_{st} \sim$ few $\times 10^{17}GeV$, while $< 45_H > /M$ and $g_{23} < 16_H > /m$ could quite plausibly be of order $h_{33}/10$.

It is interesting to observe the symmetry properties of the a_{23} and g_{23} terms. Although $10_H \times 45_H = 10 + 120 + 320$, given that $< 45_H >$ is along B-L, which is needed to implement doublet-triplet splitting, only 120 in the decomposition contributes to the mass matrices. This contribution is however antisymmetric in the family index and, at the same time, proportional to B-L. Thus the a_{23} term fulfills the requirements of both (2) and (3) simultaneously. With only h_{ij} and a_{ij} terms however, the up and down quark mass matrices will be proportional to each other, which would yield $V_{CKM} = 1$. This is remedied by the g_{ij} coupling as follows. The 16_H has a VEV primarily along its SM singlet component transforming as to

$$U = \begin{pmatrix} 0 & \epsilon + \sigma \\ -\epsilon + \sigma & 1 \end{pmatrix} m_U, \quad D = \begin{pmatrix} 0 & \epsilon + \eta \\ -\epsilon + \sigma & 1 \end{pmatrix} m_D,$$

$$N = \begin{pmatrix} 0 & -3\epsilon + \sigma \\ 3\epsilon + \sigma & 1 \end{pmatrix} m_u, \quad L = \begin{pmatrix} 0 & -3\epsilon + \eta \\ 3\epsilon + \eta & 1 \end{pmatrix} m_D,$$

Here the matrices are multiplied by left-handed fermion fields from the left and by anti-fermion fields from the right. (U, D) stand for the mass matrices of up and down quarks, while (N, L) are the Dirac mass matrices of the neutrinos and the charged leptons.

The entries $(1, \epsilon, \sigma)$ arise respectively from the h_{33}, a_{23} and h_{23} terms in eq. (13), while η entering into D and L receives contributions from both g_{23} and h_{23}; thus $\eta \neq \sigma$. Note the quark-lepton correlations between (U, N) as well as (D, L), and the up-down correlation between (U, D) as well as (N, L). These correlations arise because of the symmetry structure of $G(224)$. The relative factor of -3 between quarks and leptons involving the ϵ entry reflects the fact that $(45_H) \propto (B-L)$, while the antisymmetry in this entry arises from the $SO(10)$ structure as explained above.

Assuming $\epsilon, \eta, \sigma \ll 1$, we obtain at the unification scale:

$$\left|\frac{m_c}{m_t}\right| \simeq |\epsilon^2 - \sigma^2|, \left|\frac{m_s}{m_b}\right| \simeq |\epsilon^2 - \eta^2|,$$

$$\left|\frac{m_\mu}{m_\tau}\right| \simeq |9\epsilon^2 - \eta^2|, |m_b| \simeq |m_\tau||1 - 8\epsilon^2|, \tag{14}$$

$$|V_{cb}| \simeq |\sigma - \eta| \approx \left|\sqrt{m_s/m_b}\left(\frac{\eta + \epsilon}{\eta - \epsilon}\right)^{1/2} - \sqrt{m_c/m_t}\left(\frac{\sigma + \epsilon}{\sigma - \epsilon}\right)^{1/2}\right|, \tag{15}$$

$$\Theta^l_{\mu\tau} \approx -3\epsilon + \eta \approx \sqrt{m_\mu/m_\tau}\left(\frac{-3\epsilon + \eta}{3\epsilon + \eta}\right)^{1/2} \tag{16}$$

The relations in eqs. (15) and (16) lead to two sum rules:

$$\left|\frac{m_b}{m_\tau}\right| \simeq \left|1 - 8\left\{\left|\frac{m_\mu}{m_\tau}\right| - \left|\frac{m_s}{m_b}\right|\right\}\right|,$$

$$\frac{m_s}{m_b} \simeq \frac{m_c}{m_t} - \frac{5}{4}V_{cb}^2 \pm V_{cb}\left[\frac{9}{16}V_{cb}^2 + \frac{1}{2}\frac{m_\mu}{m_\tau} - \frac{9}{2}\frac{m_c}{m_t}\right]^{1/2} \tag{17}$$

The superscript zero, meaning unification scale values, is not exhibited, but should be understood in all the relations in eqs. (15)–(18).

The mass matrices in eq. (14) contain 5 parameters $\epsilon, \sigma, \eta, m_D = h_{23} < 10_d >$ and $m_U = h_{33} < 10_U >$. These may be determined by using, for example, the following input values: $M_t^{phys} = 174 GeV, m_c(m_c) = 1.37 GeV, m_s (1 GeV) = 110 - 116 MeV$ and the observed masses of μ and τ. While the input value of m_s is somewhat lower than that advocated in [40], it is in good agreement with recent lattice calculations [41]. With these input values, the parameters are found to be:

$$\sigma \simeq -0.110\eta_{cb}, \ \eta \simeq -0.151\eta_{cb}, \ \epsilon \simeq 0.095\eta_\epsilon,$$

$$m_U \simeq m_t(M_U) \simeq (100 - 120) Gev,$$

$$m_D \simeq m_b(M_U) \simeq 1.5 GeV \tag{18}$$

Here η_ϵ and η_{cb} denote the phases of ϵ and V_{0cb} respectively (i.e. $\epsilon = \eta_\epsilon|\epsilon|$ etc.). We assume for simplicity that they are real (barring phase angles of $\pm 10°$). Thus, $\eta_\epsilon = \pm 1$ and $\eta_{cb} = \pm 1$. The relative signs of σ, η and ϵ get fixed by ensuring that the results are optimized as regards their agreement with observation. This yields $\eta_{cb} = \eta_\epsilon$. Note that in accord with our general expectations discussed above, each of these parameters are found to be of order $1/10$, as opposed to being $O(1)$ or $O(10^{-2})$, compared to the leading

(3,3) element. Having determined these parameters, one can now obtain the following predictions:

$$m_b(m_b) \simeq (4.6 - 4.9)GeV; V_{cb} \simeq 0.045, \tag{19}$$

$$m_{\nu_\tau}^D(M_U) \simeq m_t(M_U) \simeq 100 - 120GeV,$$

$$m_{\nu_\mu}^D(M_U) \simeq (9\epsilon^2 - \sigma^2)m_U \simeq 8GeV,$$

$$\Theta_{\mu\tau}^l \simeq -3\epsilon + \eta \simeq -0.437\eta_\epsilon (\text{for } \eta_{cb}/\eta_\epsilon = +1) \tag{20}$$

In quoting the numbers in eq. (20), we have extrapolated the GUT scale values down to low energies using the beta functions of the minimal supersymmetric extension of the Standard Model (MSSM), assuming $\alpha_s(M_Z) = 0.118$, an effective SUSY threshold of $500\,GeV$ and $tan\beta = 5$. Our results depend only weakly on these input choices, so long as $tan\beta$ is neither too large (≥ 30) not too small (≤ 2). The first two of the predictions listed above (eq. (20)) correspond to directly observed entities. The last three (eq. (21)) cannot be observed directly, but they are important because they need to be combined with the Majorana masses of the RH neutrinos to yield observable entitles (see below).

Given the bizarre pattern of quark and lepton masses and mixings, it seems remarkable that the simple pattern of fermion mass matrices, motivated by the group theory of $G(224)/SO(10)$ gives an overall fit to all of them which is good to within 10%. This includes the two successful predictions on m_b and V_{cb} (eq. (20)). It is worth noting that, in supersymmetric unified theories, the "observed" value of $m_b(m_b)$ and renormalization group studies suggest that for a wide range of the parameter $tan\beta, m_b^\circ$ should in fact be about $10 - 20\%$ lower than m_τ° [42]. This is neatly explained by the relation: $m_b^\circ \approx m_\tau^\circ(1-8\epsilon^2)$ (eq. (15)), where exact equality holds in the limit $\epsilon \to 0$ (due to $SU(4)$ color), while the decrease by $8\epsilon^2 \sim 10\%$ is precisely because the off-diagonal ϵ entry is proportional to B-L (see eq. (14)).

Specially intriguing is the result on $V_{cb} \approx 0.045$ which compares well with the observed value of $\simeq 0.04$. The suppression of V_{cb}, compared to the value of 0.17 ± 0.06 obtained from eq. (6), is now possible because the mass matrices (eq. (14)) contain an antisymmetric component $\propto \epsilon$. Such a component corrects the square-root mixing angle formula $\Theta_{sb} = \sqrt{m_s/m_b}$ (appropriate for symmetric matrices of the type given by eq. (9)) by the asymmetry factor $|(\eta + \epsilon)/(\eta - \epsilon)|^{1/2}$ (see eq. (15)), and similarly for the angle Θ_{ct}. This factor suppresses V_{cb} if η and ϵ have opposite signs. The interesting point is that, the same feature necessarily enhances the corresponding mixing angle $\Theta_{\mu\tau}^l$ in the leptonic sector, since the asymmetry factor in this case is given by $[(-3\epsilon+\eta)/(3\epsilon+\eta)]^{1/2}$ (see eq. (1)). This enhancement of $\Theta_{\mu\tau}^l$ also seems to be borne out by observation in the sense that that is a key factor in accounting for the nearly maximal oscillation angle observed at SuperK (see discussion below). Note that this intriguing correlation between the mixing angles in the quark versus leptonic sectors – that is, suppression of one implying enhancement of the other – has become possible because the ϵ-contribution is simultaneously antisymmetric and is proportional to B-L. As a result, it changes sign as one goes from the quarks to the leptons.

Taking stock, we see an overwhelming set of evidences in favor of B-L and in fact for the full $SU(4)$ color-symmetry. These include: (1) the suppression of V_{cb}, together with the enhancement of $\Theta^l_{\mu\tau}$ just mentioned above, (ii) the successful relation $m^\circ_b \approx m^\circ_\tau(1-8\epsilon^2)$, where the near equality follows from $SU(4)$ color, while the decrease of m°_b relative to m°_τ by $8\epsilon^2 \sim 10\%$ is a consequence of the (B-L)-dependence of the off-diagonal ϵ-entry, (iii) the usefulness again of the $SU(4)$ color-relation $m(\nu^\tau_{\text{Dirac}})^\circ \approx m^\circ_t$ in accounting for $m(\nu^\tau_L)$, as discussed, and (iv) the agreement of the relation $|m^\circ_s/m^\circ_\mu| = |(\epsilon^2-\eta^2)/(9\epsilon^2-\eta^2)|$ with the data, in that the ratio is naturally less than 1, if $\eta \sim \epsilon$. The presence of $9\epsilon^2$ in the denominator as opposed to ϵ^2 in the numerator is again a consequence of the off-diagonal entry being proportional to B-L. Finally, a spontaneously broken (B-L) local symmetry may well be needed to ensure preservation of baryon excess in the presence of electro-weak sphaleron effects [21].

Although all the entries for the Dirac mass matrix are now fixed, to obtain the parameters for the light neutrinos one needs to specify the Majorana mass matrix of the RH neutrinos (ν^μ_R and ν^τ_R). For concreteness, we assume that this too has the hierarchical form of eq. (9):

$$M^R_\nu = \begin{pmatrix} 0 & y \\ y & 1 \end{pmatrix} M_R \qquad (21)$$

In the spirit of our discussion that flavour symmetries are the origin of hierarchical masses, we will assume that $10^{-2} \ll |y| \leq 1/10$ as opposed to $|y|$ being ≥ 0.3 (say). A priori, $y = \eta_y|y|$ can have either sign, i.e., $\eta_y = \pm 1$. Note that Majorana mass matrices are constrained to be symmetric by Lorentz invariance. The see-saw mass matrix $(-N(M^R_\nu)^{-1}N^T)$ for the light $(\nu_\mu - \nu_\tau)$ system is then

$$M^{\text{light}}_\nu = \begin{pmatrix} 0 & A \\ A & B \end{pmatrix} \frac{m^2_U}{M_R}, \qquad (22)$$

where $A \simeq (\sigma^2 - 9\epsilon^2)/y$ and $B \simeq -(\sigma + e\epsilon)(\sigma + 3\epsilon - 2y)/y^2$. With $A \ll B$, this yields

$$m_{\nu_3} \simeq B\frac{m^2_U}{M_R}; \quad \frac{m_{\nu_2}}{m_{\nu_3}} \simeq -\frac{A^2}{B^2}; \quad tan\Theta^\nu_{\mu\tau} = \sqrt{\frac{m_{\nu_2}}{m_{\nu_3}}}, \qquad (23)$$

For a given choice of the sign of y relative to that of ϵ, and for a given mass ratio m_{ν_2}/m_{ν_3}, we can now determine y using eqs. (23) and (24), and the values of ϵ and σ obtained in eq. (19). Taking $m_{\nu_2}/m_{\nu_3} = (1/10, 1/15, 1/20, 1/30)$, the requirement of hierarchy mentioned above – i.e. $10^{-2} \ll |y| \leq 0.2$ (say) – can be satisfied only provided y is positive relative to ϵ, i.e., $\eta_y = \eta_\epsilon$; corresponding values for y are: $y = (0.0543, 0.0500, 0.0468, 0.0444, 0.0424)\eta_\epsilon$. With $\eta_y = \eta_\epsilon = \pm 1$, we obtain for the neutrino oscillation angle:

$$\Theta^{osc}_{\nu_\mu \nu_\tau} \simeq \Theta^l_{\mu\tau} - \Theta^\nu_{\mu\tau} \simeq \left(0.437 + \sqrt{\frac{m_{\nu_2}}{m_{\nu_3}}} \right) (-\eta_\epsilon) \tag{24}$$

$$sin^2 2\Theta^{osc}_{\nu_\mu \nu_\tau} = (0.96, 0.91, 0.86, 0.83, 0.81)$$

$$\text{for} \quad m_{\nu_2}/m_{\nu_3} = (1/10, 1/15, 1/20, 1/25, 1/30) \tag{25}$$

Note the interesting point that just the requirement that $|y|$ should have a natural hierarchical value leads to $\eta_y = \eta_\epsilon$, and that in turn implies that the two contributions in eq. (25) must add rather than subtract, leading to an almost maximal oscillation angle. The other factor contributing to the enhancement of $\Theta^{osc}_{\nu_\mu \nu_\tau}$ is, of course, also the asymmetry-ratio which increased $\Theta^l_{\mu\tau}|$ from 0.25 to 0.437 (see eqs. (17) and (21)). We see that one can derive rather plausibly a large $\nu_\mu - \nu_\tau$ oscillation angle $sin^2 2\Theta^{osc}_{\nu_\mu \nu_\tau} \geq 0.8$, together with an understanding of hierarchical masses and mixings of the quarks and the charged leptons, while maintaining a large hierarchy in the see-saw derived masses ($m_{\nu_2}/m_{nu_3} = 1/10 - 1/30$) of ν_μ and ν_τ, all within a unified framework including both quarks and leptons. In the example exhibited here, the mixing angles for the mass eigenstates of neither the neutrinos nor the charged leptons are really large, $\Theta^l_{\mu\tau} \simeq 0.437 \simeq 23°$ and $\Theta^\nu_{\mu\tau} \simeq (0.18 - 0.31) \approx (10 - 18)°$, yet the oscillation angle obtained by combining the two is near-maximal. This contrasts with most previous work, in which a large oscillation angle is obtained either entirely from the neutrino sector (with nearly degenerate neutrinos) or almost entirely from the charged lepton sector.

It is worth noting that the interplay due to the mixing in the Dirac and the Majorana mass matrices via the see-saw mechanism has the net effect of enhancing $M_R \approx B(m_{\nu_2}/m_{\nu_e})$ for a given m_{ν_3} precisely by a factor of $|B| \approx 5$ (see eq. (23)), compared to what it would be without mixing. Using $m_U \approx 100 GeV$ (see eq. (7) or (19)) $m_{\nu_3} \approx (1/10 - 1/30)eV$ (SuperK result) and $|B| \approx 5$, one gets:

$$M_R \approx (5 - 15) \times 10^{14} GeV \tag{26}$$

Compare this with its counterpart, estimated in eq. (6), which yields $M_{3R} \approx$ few $\times 10^{14} GeV$, for $f_{33}\eta^2 \approx 1$, if $M \approx M_{\text{Planck}}$. It is interesting that the larger value of $M_R \approx 10^{15} GeV$ goes well with the theoretical estimate of eq. (6) if the characteristic mass M is chosen (perhaps more appropriately) to be $M_{\text{string}} \approx 4 \times 10^{17} GeV$ rather than M_{Planck}. Further, this larger value of M_R also goes well with the observed m_{ν_3}, once one includes the effect of mixing.

Inclusion of the first family: The first family may now be included following the spirit of the hierarchical structure shown in eqs. (9) and (14). As mentioned before, this may have its origin in flavor symmetries of a deeper theory. In the absence of such a deeper understanding, however, the theoretical uncertainties in dealing with the masses and mixings of the first family are much greater than for the heavier families, simply because the masses of the first family are so small that relatively small perturbations can significantly affect their values.

Assuming that flavor symmetries and $SO(10)$ permit the (3, 3) coupling at a genuine cubic level, but the (2, 3) couplings only at the quartic level, which are thus effectively suppressed by about an order of magnitude compared to the (3, 3) element (see discussion following eq. (13)), we would naturally expect that the (1, 2) and (1, 3) couplings (e.g., a_{12} and g_{12}, see below) would be suppressed compared to the corresponding (23) couplings. This in turn would account for the observed inter-family mass hierarchy.

Following this as a guide, and in the interest of economy, we add only two effective quartic couplings to eq. (13) to include the first family: $a_{12}16_116_2$ $45_H10_H/M$ and $g_{12}16_116_216_H16_H/M$. The first coupling introduces an ϵ' term in the (1, 2) entry, which is antisymmetric and proportional to B-L (analog of ϵ); the second introduces an η' term in the (1, 2) entry of only D and L, which is symmetric. The resulting 3×3 Dirac mass matrices are:

$$U = \begin{pmatrix} 0 & \epsilon' & 0 \\ -\epsilon' & 0 & \epsilon + \sigma \\ 0 & -\epsilon + \sigma & 1 \end{pmatrix} m_U,$$

$$D = \begin{pmatrix} OY & \epsilon' + \eta' & 0 \\ -\epsilon' + \eta' & 0 & \epsilon + \eta \\ 0 & -\epsilon + \eta & 1 \end{pmatrix} m_D,$$

$$N = \begin{pmatrix} 0 & -3\epsilon' & 0 \\ 3\epsilon' & 0 & -3\epsilon + \sigma \\ 0 & 3\epsilon + \sigma & 1 \end{pmatrix} m_U,$$

$$L = \begin{pmatrix} 0 & -3\epsilon' + \eta' & 0 \\ 3\epsilon' + \eta' & 0 & -3\epsilon + \eta \\ 0 & 3\epsilon + \eta & 1 \end{pmatrix} m_D \tag{27}$$

With $\epsilon, \sigma, \eta, m_U$ and m_D determined essentially be considerations of the second and the third families (eq. (19)), we now have just two new parameters in eq. (28), i.e., ϵ' and η' which describe five new observables in the quark and charged lepton sector: m_u, m_d, e_e, Θ_C and V_{ub}. Thus with $m_u \approx 1.5 MeV$ (at M_U) and m_e/m_μ taken as inputs one obtains: $e^1 \simeq \sqrt{m_{mu}/m_c}(m_c/m_t) \approx 2 \times 10^{-4}$ and $|\eta'| \simeq \sqrt{m_e/m_\mu}(m_\mu/m_\tau) \simeq 4.4 \times 10^{-3}$. We can now calculate m_d, Θ_c and V_{ub}. Combining the two predictions for the second and the third families obtained before (see eq. (19)), we are thus led to a total of five predictions for the observable parameters of the quarks and charged leptons belonging to the three families.

$$m_b(m_b) \quad \simeq \quad (4.6 - 4.9)GeV$$

$$V_{cb} \quad \simeq \quad 0.045$$

$$m_d(1GeV) \quad \simeq \quad 8MeV$$

$$\Theta_C \quad \simeq \quad |\sqrt{m_d/m_s} - c^{\iota\phi}\sqrt{m_u/m_c}|$$

$$|V_{ub}/V_{cb}| \quad \simeq \quad \sqrt{m_\mu/m_c} \simeq 0.07 \tag{28}$$

Further, the Dirac masses and mixing of the neutrinos and the mixings of the charged leptons also get determined. Including those for the $\mu - \tau$ families listed in eq. (21), we obtain:

$$M_{\nu_\tau}^D \approx 100 - 120 GeV'm_{\nu_\mu}^D(M_U) \simeq 8Gev, \Theta_{\mu\tau}^l \simeq -0.437\eta_\epsilon,$$

$$m_{\nu_\epsilon}^D \simeq [9\epsilon'^2/(9\epsilon^2 - \sigma^2)]m_U \simeq 0.4MeV,$$

$$\Theta_{\epsilon\mu}^l \simeq \left[\frac{\eta' - e\epsilon'}{\eta' + e\epsilon'}\right]^{1/2} \sqrt{m_e/m^\mu} \simeq 0.85\sqrt{m_e/m_\mu} \simeq 0.06,$$

$$\Theta_{\epsilon\tau}^l \simeq \frac{1}{0.85}\sqrt{m_e/m_\tau}(m_u/m_\tau) \simeq 0.0012. \tag{29}$$

In evaluating $\Theta_{\epsilon\mu}^l$, we have assumed ϵ' and η' to be relatively positive.

Note that the first five predictions in eq. (29) pertaining to observed parameters in the quark system are fairly successful. Considering the bizarre pattern of the masses and mixings of the fermions in the three families (recall comments on $V_{cb}, m_b/m_\tau, m_s/m_\mu$ and m_d/m_e), we feel that the success of the mass pattern exhibited by eq. (28) is rather remarkable. This is one reason for taking patterns like eq. (28) seriously as a guide for considerations on proton decay. A particularly interesting variant is obtained in the limit $\epsilon' \to 0$, as I will mention later.

To obtain some guidelines for the neutrino system involving ν_e, we need to extend the Majorana mass matrix of eq. (22), by including entries for ν_R^e. Guided by economy and the assumption of hierarchy, as in eq. (9), we consider the following pattern:

$$M_\nu^R = \begin{pmatrix} x & 0 & 1 \\ 0 & 0 & y \\ z & y & 1 \end{pmatrix} M_R \tag{30}$$

Equation (30) introduces four effective parameters: x, y, z and M_R. The magnitude of $M_R \approx (5 - 50) \times 10^{14}GeV$ can quite plausibly be justified in the context of supersymmetric unification (see estimate given in eq. (6) and discussion following eq. (27)). And, to the same extent, the magnitude of $m(\nu_\tau) \approx (1/10 - 1/30)eV$, which is consistent with the SuperK value, can also be anticipated. Since all the Dirac parameters are determined, there are, effectively, three new parameters: $x, y,$ and z. However, there are six observables in the light three neutrino system: the three masses and the three oscillation angles. Thus one can expect three predictions for the light neutrinos. These may be taken to be $\Theta_{\nu_\mu\nu_\tau}^{osc}$ (eq. (25)), m_ν, (see eqs. (8) and (24)), and for example, $\Theta_{\nu_e\nu_\mu}^{osc}$.

Recall that the parameter y was determined above by assuming that the MSW (small or large angle) solution for the solar neutrino-deficit corresponds to $\nu_e - \nu_\mu$ oscillation, with $(\delta m^2)MSW \approx m(\nu_\mu)^2 \sim 10^{-5}eV^2$. This gave a value of $|y| \approx 1/20$, in full accord with our general expectation of a hierarchy of order $(1/10)$ for the $(2, 3)$ entry compared to the $(3, 3)$. We do not, however, have much experimental information at present, to determine the other two parameters x and y, reliably, because very little is known about the observable parameters involving ν_e. To have a feel, consistent with our presumption that the inter-family hierarchical masses arise through successively smaller off-diagonal mixing elements, we will assume that $y \approx 1/20$ (as above), $z \leq y/10$ and $x \sim z^2$. Thus, in addition to $M_R \approx (5 - 15) \times 10^{14} GeV$ and $y \approx 1/20$, which as mentioned above are better determined, we take as a guide: $z \sim (1 - 5) \times 10^{-3}$ and $x \sim (1 \text{ to few}) (10^{-6} - 10^{-5})$. Including the three predictions mentioned above, the mass eigenvalues and the oscillation angles are then:

$$m_{\nu\tau} \approx (1/10 - 1/30)eV$$

$$m_{\nu\mu} \simeq 10^{-3}(5 \text{ to } 1)eV$$

$$m_{\nu e} \simeq (10^{-5} - 10^{-4})(1 \text{ to } \text{ few})eV$$

$$\Theta^{osc}_{\mu\tau} \simeq 0.437 + \sqrt{m_{\nu_2}/m_{\nu_3}}$$

$$\Theta^{osc}_{e\mu} \simeq \Theta^l_{e\mu} - \Theta^\nu_{e\mu} \simeq 0.06 \pm 0.015$$

$$\Theta^{osc}_{e\tau} \simeq \Theta^l_{e\tau} - \Theta^\nu_{e\tau} \simeq 10^{-3} \pm 0.03 \tag{31}$$

We see that the masses of ν_e and ν_μ and the oscillation angle $\Theta^{osc}_{e\mu}$ goes well with the small angle MSW explanation of the solar neutrino-deficit.

Although, the superheavy Majorana masses of the RH neutrinos cannot be observed directly, they can be of cosmological significance. The pattern given earlier and in this section suggests that $M(\nu^\tau_R) \approx (5-15) \times 10^{14} GeV$, $M(\nu^\mu_R) \approx (1 - 4) \times 10^{12} GeV$ (for $y \approx 1/20$); and $M(\nu^e_R) \sim (1/2 - 10) \times 10^9 GeV$ (for $x \sim (1/2 - 10)10^{-6} > z^2$). A mass of $\nu^e_R \sim 10^9 GeV$ is of the right magnitude for producing ν^e_R following reheating and inducing lepton asymmetry in ν^e_R decay into $H^\circ + \nu^i_L$, that is subsequently converted into baryon asymmetry by the electro-weak sphalerons [21].

We have demonstrated that a rather simple pattern for the four Dirac mass matrices, motivated and constrained by the group structure of $SO(10)$, is consistent within 10% with the observed masses and mixing of all the quarks and the charged leptons. This fit is significantly over constrained, leading to five predictions, which are successful. The same pattern, supplemented with a similar structure for the Majorana mass matrix, quite plausibly accounts for the SuperKamiokande result with the large $\nu_\mu - \nu_\tau$ oscillation angle required for the atmospheric neutrinos, and accommodates a small $\nu_e - \nu_\mu$ oscillation angle elevant for theories of the solar neutrino deficit.

Before turning to proton decay, it is work noting that much of our discussion of fermion masses and mixings, including those of the neutrinos, is essentially unaltered if we go to the limit $\epsilon' \to 0$ of eq. (28). This limit clearly involves:

$$m_u = 0, \Theta_C \simeq \sqrt{m_d/m_s}$$

$$|V_{ub}| \simeq \sqrt{\frac{\eta - \epsilon}{\eta + \epsilon}} \sqrt{m_d/m_b}(m_s/m_b) \simeq (2.1)(0.039)(0.023) \simeq 0.0019$$

$$m_{\nu_e} = 0, \Theta^\nu_{e\mu} = \Theta^\nu_{e\tau} = 0$$

All other predictions will remain unaltered. Now, among the observed quantities in the list above, $\Theta_C \approx \sqrt{m_d/m_s}$ is indeed a good result. Considering that $m_\mu/m_t \approx 10^{-5}, m_u = 0$ is also a pretty good result. There are of course, plausible small corrections (arising from higher dimensional operators for example), involving Planck scale physics which could induce a small value for m_u through the (1, 1) entry $\delta \approx 10^{-5}$. For considerations of proton decay, it is worth distinguishing between these two variants, which we will refer to as cases I and II respectively.

$$\text{Case I}: \quad \epsilon' \approx 2 \times 10^{-4}, \delta = 0$$

$$\text{Case II}: \quad \delta \approx 10^{-5}, \epsilon' = 0 \tag{32}$$

5 Link Between Fermion Masses and Proton Decay in Supersymmetric $SO(10)$

5.1 Preliminaries

I present now the results of a recent study [18] of proton decay in SUSY $SO(10)$, which was carried out by paying attention specially to the link that exists in SUSY $SO(10)$ between proton decay and the masses and mixings of all fermions, including especially the neutrinos.

It is well known that in supersymmetric unified theories (GUTs), with $M_X \sim 2 \times 10^{16} GeV$, the gauge-boson mediated $d = 6$ proton decay operators, for which $e^+\pi^0$ would have been the dominant mode, are strongly suppressed. The dominant mechanism for proton decay in these theories is given by effective $d = 5$ operators of the form $Q_iQ_jQ_kL_l/M$ in the superpotential, which arise through the exchange of color triplet Higgsions that are the GUT partners of the standard Higgs doublets such as those in $5_H + \bar{5}_H$ of $SU(5)$ or the 10_H of $SO(10)$. Subject to a doublet-triplet splitting mechanism which makes these color triplets acquire heavy GUT-scale masses, while the doublets remain light, these standard $d = 5$ operators, suppressed by just one power of the heavy mass and the small Yukawa couplings, lead to proton decay, with a lifetime $\tau_p \sim 10^{30} - 10^{34} yrs$ [43]–[46]. Note that these standard $d = 5$ operators are proportional to the product of two Yukawa couplings, which are related to the masses and mixing of the charged fermions. Further, for these operators to induce proton decay, they must be dressed by wino (or gluino)-exchange so as to convert a pair of suqrks to quarks. Owing to (a) Bose symmetry of the superfields in QQQL/M, (b) color antisymmetry, and especially (c) the hierarchical Yukawa couplings of the standard Higgs doublets, it turns out that these operators exhibit a strong preference for the decay of a proton into channels involving $\bar{\nu}$ rather than e^+ or (even) μ^+ and those involving an $\bar{s}$ rather than a $\bar{d}$. Thus the standard operators lead to dominant $\bar{\nu} K^+$ and comparable $\bar{\nu} K^+$ modes, but in all cases to highly suppressed $e^+ \pi^o, e^+ K^o, e^+ K^o$ and even $\mu^+ K^o$ modes. For instance, for SUSY $SU(5)$ one obtains (for $tan\beta \leq 15$, say):

$$[\Gamma(\mu^+ K^o)/\Gamma(\bar{\nu}_\mu K^+)]_{std} \sim [m_u/m_c sin^2\Theta_C]^2 R \approx 10^{-3}$$

where $R \approx 0.1$ is the ratio of the products of the relevant $|$ matrix element $|^2 \times$ (phase space) for the two modes.

Now, it was recently realized that in left-right symmetric unified theories possing super-symmetry, such as those based on $G(224)$ or $SO(10)$, there is very likely a new source of $d = 5$ proton decay operators, which are related to the Majorana masses of the right-handed neutrinos [17]. For instance, in the context of the minimal set of Higgs multiplets $\{45_H, 16_H, \bar{16}_H$ and $10_H\}$, which have been utilized earlier to break $SO(10)$ and generate fermion masses, these new $d = 5$ operators arise by combining three effective couplings, i.e., (a) the couplings $f_{ij}16_i 16_j \bar{16}_H \bar{16}_H/M$ (see eq. (5)) which are essential to assign Majorana masses to the right-handed neutrinos, (b) the couplings $g_{ij}16_i 16_j 16_H 16_H/M$, which are needed to generate non-trivial CKM mixing and (c) the mass term $M_{16}16_H \bar{16}_H$. In the presence of these three (unavoidable) effective couplings and the VEVs $<16_H> = <\bar{16}_H> \sim M_x$, the color triplet Higgsinos in 16_H and $\bar{16}_H$ of mass M_{16} cab be exchanged between $\bar{q}_i q_j$ and $\bar{q}_k l_l$ pairs. This exchange gives rise to a new set of effective $d = 5$ couplings of the form:

$$L_{new}^{d=5}[f_{ij} gkl(16_i 16_j)(16_k 16_t)/M_{16}]\frac{<\bar{16}_H><16_H>}{M^2} \tag{33}$$

which induce proton decay, just as the standard operators do. Note that these new $d = 5$ operators depend, through the couplings f_{ij} and g_{kl}, both on the Majorana and on the Dirac masses of the respective fermions. This is why within SUSY $G(224)$ or $SO(10)$, proton decay gets intimately linked to the masses and mixings of all fermions, including neutrinos.

Specifically, it is found that the SuperK result on atmospheric neutrinos, that suggests $m(\nu_L^\tau) \sim 1/20 eV$ and a large $\nu_\mu - \nu_\tau$ oscillation angle leads to a significant enhancement especially in the new $d = 5$ operators, compared to previous estimate which were based on guesses of much larger values of $m(\nu_L^\tau) \sim (2 - 4)eV$ [17]. Curiously enough, the net effect of including the enhancement of f_{33} (due to a lowering of $m(\nu_L^\tau)$) and the suppression of the relevant CKM mixings is such that the strength of the new $d = 5$ operators is found to be comparable to that of the standard ones [18]. The flavor structure of the new operators are, however, very different from those of the standard ones, in part because the former depend on the Majorana masses of the RH neutrinos, and the latter do not. As a result, the new operators lead to some characteristic differences in the proton decay pattern (that is, branching ratios of different decay modes) compared to the standard ones (see below).

5.2 Framework for Calculating Proton Decay Rate

To establish notations, consider the case of minimal SUSY $SU(5)$ and, as an example, the process $\bar{c}\bar{d} \to \bar{s}\bar{\nu}_\mu$, which induces $p \to \bar{\nu}_\mu K^+$. Let the strength of the corresponding $d = 5$ operator, multiplied by the product of the CKM mixing elements entering into wino-exchange vertices, (which in this case is $sin\Theta_C cos\Theta_C$) be denoted by $\bar{A}$. Thus, putting $cos\Theta_C = 1$, one obtains:

$$\bar{A}_{\bar{c}\bar{d}}(SU(5)) = (h^\mu_{22} h^d_{12}/M_{H_c})sin\Theta_c \simeq (m_c m_s sin^2\Theta_C/v_u^2)(tan\beta/M_{H_c}$$

$$\simeq (1.9 \times 10^{-8})(tan\beta/M_{H_c}, \tag{34}$$

where $tan\beta \equiv v_u/v_d$, and we have put $u_u = 174 GeV$ and the fermion masses extrapolated to the unification scale, i.e., $m_c \simeq 300 MeV$ and $m_s \simeq 40 MeV$. The amplitude for the associated four fermion process $dus \to \bar{\nu}$ is given by:

$$A_5(dus \to \bar{\nu}_\mu) = (\bar{A}_{\bar{c}\bar{d}}) \times (2f), \tag{35}$$

where f is the loop factor associated with wino-dressing. Assuming $m_{\bar{w}} \ll m_{\bar{q}} \sim m_j$ one gets; $f \simeq (m_{\bar{w}}/m_{\bar{q}}^2)(\alpha_2/4\pi)$. Using the amplitude for $(du)(s\nu_l)$, as in eq. (35), $(l = \mu \, or \, \tau)$, one then obtains [44]–[46], [18]:

$$\Gamma^{-1}(p \to \bar{\nu}_\tau K^+) \simeq (2.2 \times 10^{31})yrs \times \left[\frac{0.67}{A_S}\right]^2 \left[\frac{0.006 GeV^3}{\beta_H}\right]^2$$

$$\left[\frac{(1/6)}{(m_{\bar{w}}/m_{\bar{q}})}\right]^2 \left[\frac{m_{\bar{q}}}{1 TeV}\right]^2 \left[\frac{2 \times 10^{-24} GeV^{-1}}{\hat{A}(\bar{\nu})}\right]^2 \tag{36}$$

Here β_H denotes the hadronic matrix element defined by $\beta_H u_L(\boldsymbol{k})(\boldsymbol{t} \equiv \epsilon_{\alpha\beta\gamma} \langle 0|(d_L^\alpha u_L^\beta)u_L^\gamma|p, \boldsymbol{k}\rangle$. While the range $\beta_H = (0.003 - 0.03)GeV^3$ has been used in the past [45], given that one lattice calculations yield [50] $\beta_H = (5.6 \pm 0.5) \times 10^{-3} GeV^3$, we will take as a plausible range: $\beta_H = (0.006 GeV^3)(1/2$ to 2). $A_s \approx 0.67$ stands for the short distance renormalization factor of the $d = 5$ operator. Note that the familiar factors that appear in the expression for proton lifetime – i.e., M_{H_C}, $(1 + y_t K)$ representing the interference between the $\bar{t}$ and $\bar{c}$ contributions and $tan\beta$ – are all effectively contained in $\hat{A}(\bar{\nu})$. Allowing for plausible and rather generous uncertainties in the matrix element and the spectrum we take:

$$\beta_H = (0.0006 GeV^3)(1/2 \, to \, 2),$$

$$(m_{\tilde{w}}/m_{\tilde{q}}) = 1/6(12\,\text{to}\,2),\, m_{\tilde{q}} \approx m_{\tilde{l}} \approx 1TeV(1/\sqrt{2}\,\text{to}\,\sqrt{2}) \tag{37}$$

Using eqs. (36) and (37), we get:

$$\Gamma^{-1}(p \to \nu_\tau K^+) \approx (2.2 \times 10^{31})yrs$$

$$\times\, [2.2 \times 10^{-24}GeV^{-1}/\hat{A}(\bar{\nu}_l)]^2[32\,\text{to}\,/32] \tag{38}$$

This relation is general, depend only on $\hat{A}(\bar{\nu}_l)$ and on the range of parameters given in eq. (38). It can thus be used for both $SU(5)$ and $SO(10)$.

The experimental lower limit on the inverse rate for the $\bar{\nu}K^+$ modes is given by [47],

$$[\sum_l \Gamma(p \to \bar{\nu}_l K^+)]^{-1}_{expt} > 7 \times 10^{32}yrs. \tag{39}$$

Allowing for all the uncertainties to stretch in the same direction (in this case, the square bracket = 45), and assuming that just one neutrino flavor (e.g. ν_μ for $SU(5)$) dominates, the observed limit (eq. (40)) provides an upper bound on the amplitude:

$$\hat{A}(\bar{\nu}_l) \leq 2 \times 10^{-24}GeV^{-1}, \tag{40}$$

which holds for both $SU(5)$ and $SO(10)$. For minimal $SU(5)$, using eq. (35) and $tan\beta \geq 2$ (which is suggested on several grounds), one obtains a lower limit on M_{HC} given by:

$$M_{HC} \geq 2 \times 10^{16}GeV(SU(5)) \tag{41}$$

At the same time, higher values of $M_{HC} > 3 \times 10^{16}GeV$ do not go very well with gauge coupling unification [48]. Thus, keeping $M_{HC} \leq 3 \times 10^{16}$ and $tan\beta \leq 2$, we obtain from eq. (35):

$$\hat{A}(SU(5)) \geq (4/3) \times 10^{-24}GeV^{-1}$$

Using eq. (39), this in turn implies that

$$\Gamma^{-1}(p \to \bar{\nu}K^+) \leq 1.5 \times 10^{33}yrs(SU(5)) \tag{42}$$

This a conservative upper limit. In practice, it is unlikely that all the uncertainties, including that in M_{HC}, would stretch in the same direction to nearly extreme values so as to prolong proton lifetime. A more reasonable upper limit, for minimal $SU(5)$, thus seems to be:

$$\Gamma^{-1}(p \to \bar{\nu}K^+)(SU(5)) \leq (0.7) \times 10^{33}yrs.$$

Given the experimental lower limit (eq. (40)), we see that minimal SUSY $SU(5)$ is almost on the verge of being excluded by proton decay searches. We have of course noted earlier that SUSY $SU(5)$ does not go well with the neutrino oscillations observed at SuperK.

Now, to discuss proton decay in the context of supersymmetric $SO(10)$, it is necessary to discuss first the mechanism for doublet-triplet splitting. Details of this discussion may be found in [18]. Here, I present only a synopsis.

5.3 A Natural Doublet-Triplet Splitting Mechanism in $SO(10)$

In supersymmetric $SO(10)$, a natural doublet-triplet splitting can be achieved by coupling the adjoint Higgs 45_H to a 10_H and a $10'_H$ with 45_H acquiring a unification scale VEV in the B-L direction [49]: $\langle 45_H \rangle = (a, a, a, 0, 0) \times \tau_2$ with $a \sim M_U$. As discussed already, to generate CKM mixing for fermions, we require an $\langle 16_H \rangle_d$ to acquire an electro-weak scale vacuum expectation value. To insure accurate gauge coupling unification, the effective low energy theory should not contain split multiplets beyond those of MSSM. Thus the MSSM Higgs doublets must be linear combinations of the $SU(2)_L$ doublets in 10_H and 16_H. A simple set of superpotential terms that ensures this and incorporates doublets in 10_H and 16_H. A simple set of superpotential terms that ensures this and incorporates doublet-triplet splitting is:

$$W_H = \lambda 10_H 45_H 10'_H + M_{10} 10'^2_H + \lambda' \overline{16_H 16_H} 10_H + M_{10} 16_H \bar{16}_H \qquad (43)$$

A complete superpotential for $45_H, 16_H, \bar{16}_H, 10_H, 10'_H$ and possibly other fields which ensure that $45_H, 16_H$, and $\bar{16}_H$ acquire unification scale VEVs with $\langle 45_{H} \rangle$ being along the (B-L) direction, that exactly two Higgs doublets (H_u, H_d) remain light with H_d being a linear combination of $(10_H)_d$ and $(16_H)_d$, and that there are no unwanted pseudoGoldstone bosons, can be constructed with the vacuum expectation value $\langle 45_H \rangle$ in the B-L direction. It does not contribute to the doublet matrix, so one pair of Higgs doublet remains light, while all triplets acquire unification scale masses. The light MSSM Higgs doublets are

$$H_u = 10_u, H_d = cos_\gamma 10_d + sin_\gamma 16_d, \qquad (44)$$

with $tan\gamma \equiv \lambda' \langle \bar{16}_H \rangle / M_{16}$. Consequently, $\langle 10 \rangle_d = cos\gamma v_d, \langle 16_d \rangle = sin\gamma v_d$ with $\langle H_d \rangle = v_d$ and $\langle 16_d \rangle$ and $\langle 10_d \rangle$ denoting the electro-weak VEVs of those multiplets. Note that the H_u is purely in 10_H and that $\langle 10_d \rangle^2 + \langle 16_d \rangle^2 = v_d^2$. This mechanism of doublet-triplet (DT) splitting is rather unique for the minimal Higgs systems in that it meets the requirements of both D-T splitting and CKM mixing. In turn, it has three important consequences:

(i) It modifies the familiar $SO(10)$ relation $tan\beta \equiv v_u/v_d = m_t/m_b \approx 60$ to

$$tan\beta/cos\gamma \approx m_t/m_b \approx 60 \qquad (45)$$

As a result, even low to moderate values of $tan\beta \approx 3$ to 10 (say), are perfectly allowed in $SO(10)$ (corresponding to $cos\gamma \approx 1/20$ to $1/6$).

(ii) In contrast to $SU(5)$, for which the strengths of the standard $d = 5$ operators are proportional to $(M_{H_c}^{-1}, M_{H_c} \sim M_U \sim$ few $\times 10^{16} GeV$ (see eq. (35)), for the $SO(10)$ model, with DT splitting given as above, they become proportional to M_{eff}^{-1}, where $M_{eff} = (\lambda a)^2/M_{10'} \sim M_U^2/M_{10'}.M_{10'}$ can be naturally smaller than M_U, and thus M_{eff} is correspondingly larger (than M_U) by one or two orders of magnitude [18]. Now, the proton decay amplitudes for $SO(10)$

in fact possess an intrinsic enhancement compared to those for $SU(5)$, owing primarily due to differences in their Yukawa couplings for the up sector (see Appendix C of [18]). As a result, these larger values of $M_{eff} \sim 10^{18} GeV$ are found to lead to expected proton decay lifetimes that are on the one hand compatible with observed limits, but on the other hand allow optimism as regards future observation of proton decay (see below).

(iii) M_{eff} gets bounded above by considerations of coupling unificatioin and GUT scale threshold effects. Owing to mixing between 10_d and 16_d (see eq. (45)), the correction to $\alpha_3(m_z)$ due to doublet-triplet splitting becomes proportional to $ln(M_{eff}/cos\gamma)$. Inclusion of this correction and those due to splittings within the gauge multiplets (i.e. 45_H, and $\overline{16}_H$), together with the observed degree of coupling unification allows us to obtain a conservative upper limit on $M_{eff} \leq 3 \times 10^{18} GeV$ (see [18]). This in turn helps provide an upper limit on the expected proton decay lifetime (see below):

The calculation of the amplitudes $\hat{A}_{std}$ and $\hat{A}_{new}$ for the standard and the new operators for the $SO(10)$ model are given in detail in [18]. Here, I will present only the results. It is found that the four amplitudes $\hat{A}_{std}(\bar{\nu}_\tau K^+)$, $\hat{A}_{std}(\bar{\nu}_\mu K^+)$, $\hat{A}_{new}(\bar{\nu}+)$ and $\hat{A}_{new}(\bar{\nu}_\nu K^+)$ are in fact very comparable to each other, within about a factor of two, either way. Since there is no reason to expect a near cancellation between the standard and the new operators, especially for both $\bar{\nu}_\tau K^+$ and $\bar{\nu}_\mu K^+$ modes, we expect the net amplitude (standard + new) to be in the range exhibited by either one. Following [18], I therefore present the contributions from the standard and the new operators separately. Using the upper limit on $M_{eff} \leq 3 \times 10^{18} GeV$, we obtain a lower limit for the standard proton decay amplitude given by

$$\hat{A}(\bar{\nu}_\tau K^+)_{std} \geq \begin{bmatrix} (7 \times 10^{-24} GeV^{-1})(1/6 \,\text{to}\, 1/4) \\ (3 \times 10^{-24} GeV^{-1})(1/6 \,\text{to}\, 1/2) \end{bmatrix} \tag{46}$$

Substituting into eq. (39) and adding the contribution from the second competing mode, $\bar{\nu}_\mu K^+$ with a typical branching ratio $R \approx 0.3$, we obtain

$$\Gamma^{-1}(\bar{\nu} K^+)_{std} \leq \begin{bmatrix} (3 \times 10^{31} yrs.)(1.6 \,\text{to}\, 0.7) \\ (6.8 \times 10^{31} yrs.)(4 \,\text{to}\, 0.44) \end{bmatrix} (32 \,\text{to}\, 1/32) \tag{47}$$

The upper and lower entries in eqs. (47) and (48) henceforth correspond to the cases I and II of the fermion mass matrix (i.e., $\epsilon' \nu 0$ and $\epsilon' = 0$, respectively, see eq. (33)). The uncertainty shown inside the square brackets correspond to that in the relative phases of the different contributions. The uncertainty $(32 \,\text{to}\, 1/32)$ corresponds to the uncertainty in β_H, $(m_{\bar{W}}/m_{\bar{q}})$ and $m_{\bar{q}}$, by factors of 2, 2, and $\sqrt{2}$ respectively, either way, around the "central" values reflected in eq. (38). Thus, we find that for MSSM embedded in $SO(10)$, the inverse partical proton decay rate should satisfy:

$$\Gamma^{-1}(p \to \bar{\nu} K^+)_{std} \leq \begin{bmatrix} 3 \times 10^{31} \pm 1.7 yrs. \\ 6.8 \times 10^{31+2.1}_{-1.5} yrs. \end{bmatrix}$$

$$\leq \begin{bmatrix} 1.5 \times 10^{33} yrs. \\ 7 \times 10^{33} yrs. \end{bmatrix} (SO(10)) \tag{48}$$

The central value of the upper limit in eq. (49) essentially reflects the upper limit on M_{eff}, while the remaining uncertainties of matrix elements and spectrum are reflected in the exponents.

Evaluating similarly the contribution from the new operator, we obtain:

$$\hat{A}(\bar{\nu}_\mu K^+)_{new} \approx (1.5 \times 10^{-24} GeV^{-1})(1/4 \text{ to } 1.3) \tag{49}$$

$$\Gamma^{-1}(\bar{\nu}K^+)_{new} \approx (3 \times 10^{31} yrs)[16 \text{ to } 1/1, 7]\{32 \text{ to } 1/32\} \tag{50}$$

In this estimate we have included the contribution of the $\bar{\nu}_\tau K^+$ mode with a typical branching ratio $R \approx 0.4$. Here the second factor, inside the square bracket, reflects the uncertainties in the amplitude, while the last factor corresponds to varying $\beta_H, (m_{\bar{W}}/m_{\bar{d}}$ and $m_{\bar{q}}$ around the central values reflected in eq. (38). With a net factor of even 20 to 100 arising jointly from the square and the curly brackets, i.e. without going to extreme ends of all parameters, the new operators related to neutrino masses lead by themselves to proton decay lifetimes

$$\Gamma^{-1}(\bar{\nu}K^+)_{new}^{expected} \approx (0.6 - 3) \times 10^{33} yrs.(SO(10)) \tag{51}$$

The Charged Lepton Decay Mode $(p \to \mu^+ K^\circ)$: I now discuss a special feature of the $SO(10)$ model pertaining to the possible prominence of te charged lepton decay mode: $p \to \mu^+ K^\circ$, which is not permissible in SUSY $SU(5)$. Allowing for uncertainties in the way the standard and the new operators can combine with each other for the three leading modes, i.e., $\bar{\nu}_\tau K^+, \bar{\nu}_\mu K^+$ and $\mu^+ K^\circ$, we obtain [18]

$$B(\mu^+ K^\circ)_{std} + new \approx [1 \text{ to } (50 - 60)\%]\rho(SO(10)) \tag{52}$$

where ρ denote the ratio of the squares of relevant matrix elements for the $\mu^+ K^\circ$ and $\bar{\nu}K^+$ modes.

In the absence – presumably temporary – of a reliable lattice calculation, which is presently missing for the $\bar{\nu}K^+$ mode [50], one should remain open to the possibility of $\rho \approx 1/2$ to 1 (say). Using eq. (53), we find that for a large range of parameters, the branching ratio $B(\mu^+ K^\circ)$ can lie int he range of 20–30% (if $\rho \approx 1$). Thus we see that the $\mu^+ K^\circ$ mode is likely to be prominent in the $SO(10)$ model presented here, and if $\rho \approx 1$, it can even become a dominant mode. This contrasts sharply with the minimal $SU(5)$ model in which the $\mu^+ K^\circ$ is expected to have a branching ratio of only about 10^{-3}. In the $SO(10)$ model, the standard operator by itself gives a branching ratio for this mode of (1–10)% while the potential prominence of the $\mu^+ K^\circ$ mode arises only through the new operator related to neutrino masses.

6 Some Crucial Observations Pertaining to Unification: A Summary

The preceding discussion can be best summarized by listing the implications of some crucial findings which bear on unification.

A. The family multiplet structure: The observed multiplet structure in each family consisting of either sixteen members (including the ν_R) or fifteen members (without ν_R) is the first empirical hint in favor of an underlying gauge symmetry like $G(224), SO(10)$ or $SU(5)$. While the standard model organizes the 15 members of a family into five multiplets, $SU(5)$ groups them into two, and $G(224)$ with L-R discrete symmetry, $SO(10)$ places all sixteen members within just one multiplet. Further, each of these higher symmetries $(G(224), SO(10)$ or $SU(5))$ explain precisely the $SU(3)^C \times SU(2)_D$-representations and the weak hypercharge (Y_W), quantum numbers of all the members in a family. This feature as well as the need to explain the observed quantization of electric charge, have been two of the primary motivations for proposing the idea of grand unification [5]–[7].

B. Meeting of the gauge couplings: The meetings of the gauge couplings, which is found to occur when their measured values at LEP are extrapolated to higher energies in the context of supersymmetry, clearly supports the ideas of:

• An underlying unity of forces, as well as of supersymmetry
• The relevance of effective gauge symmetries like $SU(5)$, or $SO(10)$, or a string-derived $G(224)$, or $[SU(3)]^3$ at the underlying level
• Unification at a scale $M_x \sim 2 \times 10^{16} GeV$ (assuming MSSM spectrum below M_x)

C. Neutrino masses, especially m(ν_τ) $\sim$ 1/20eV: This single piece of information, suggested by the SuperK result, brings to light the existence of the RH neutrinos accompanying the left-handed ones, and reinforces the ideas of:

• SU(4) color
• Left-right symmetry
• Supersymmetric unification
• See-saw

In short, the SuperK result, suggesting $m(\nu_\tau) \sim (1/20)eV$, slects out the route to higher unification based on a string-derived $G(224)$ or $SO(10)$, as opposed to $SU(5)$. Further, it suggests that B-L breaking occurs at the unification scale, $M_{B-L} \sim M_X \sim 2 \times 10^{10} GeV$ rather than at an intermediate scale.

D. Masses and mixings of all fermions (q, l, ν): Adopting familiar ideas of generating lighter eigenvalues through off-diagonal mixings and using the group theory of $SO(10)$ for the effective Yukawa couplings of the minimal Higgs system, it was found in [18], that, remarkably enough, the bizarre pattern of the masses and mixings of the charged fermions as well as of the neutrinos can be adequately described (with $\sim 10\%$ accuracy) within an economical and predictive $SO(10)$ framework. In particular, the framework provides five successful predictions for the masses and mixings of the quarks and the charged leptons. The same description goes extremely well with a value of $m_\nu, \sim 1/20eV$ as well as with a large $\nu_\mu - \nu_\tau$ oscillation angle $(sin^2 2\Theta_{\nu_\mu \nu_\tau} \approx 0.82 - 0.96)$, despite highly non-degenerate masses for the light neutrinos. Both these features are in good agreement with the SuperK results on atmospheric neutrinos. The same framework also typically leads to the small angle MSW solution for the solar neutrino puzzle, with $m_{\nu_e} \sim 3 \times 10^{-3}eV \gg m_{\nu_e}$.

One intriguing feature of the $SO(10)$ framework presented is that the largeness of the $\nu_\mu - \nu_\tau$ oscillation angle emerges naturally together with the smallness of the analogous mixing parameter in the quark-sector: $V_{bc} \approx 0.04$. This remarkable correlation between the leptonic versus the quark mixing angles clearly points to the presence of a contribution of the mass matrices, which is proportional to B-L, and its antisymmetric in the family space. The minimal Higgs system together with the group theory of $SO(10)$ precisely yields such a contribution.

E. Proton decay: The hall-mark of quark-lepton unification: Proton decay, if seen, would directly verify the idea of quark-lepton unification. Note that this crucial aspect of grand unification is not probed directly by the other three observations listed above: B, C, and D.

We have argued that three different sets of observations, i.e. (a) the observed meeting of the three gauge couplings, (b) the SuperK result on atmospheric neutrino oscillations, and (c) fermion masses and mixings – go extremely well with the idea of supersymmetric unification, based on symmetry structures such as $SO(10)$. Babu, Wilczek and I have studied proton decay in this context, paying attention to its correlation with fermion masses and mixings [18]. We found that the proton decay amplitudes receive a major contribution from a set of new $d = 5$ operators which are directly related to the Majorana masses of the RH neutrinos and to the CKM mixing [17, 18]. This is in addition to the contribution from the standard $d = 5$ operators, which are related to the Dirac masses of the charged fermions. The study shows that the mass of $m_{\nu_\tau} \sim 1/20eV$ (as opposed to previously considered values of a few eV) and the large oscillation angle suggested by the SuperK result, in fact imply a net enhancement in the rates of proton decay into the $\bar\nu K^+$ and especially in the $\mu^+ K^\circ$-modes [18], relative to previous estimates.

There are of course uncertainties in the prediction for proton decay rates owing to those in the SUSY spectrum, the hadronic matrix elements and the relative phases of the different contributions. Allowing for rather generous uncertainties in this regard, we expect proton to decay dominantly into the $\bar\nu K^+$ and very likely to the $\mu^+ K^\circ$-mode as well, with a lifetime:

$$\tau_{\text{proton}} \leq 7 \times 10^{33} yrs (SO(10)) \qquad (53)$$

This is a conservative upper limit which is obtained only if all the uncertainties are stretched in the same direction to nearly their extreme values, so as to extend proton longevity. Since the likelihood of this happening is small, we expect that within either a string-derived $G(224)$ or the $SO(10)$ model of the sort presented here, proton should decay with a lifetime shorter than the limit shown above. With the current experimental lower limit already at 7×10^{32} years, we conclude that improvement in the present limit for $p \to \bar{\nu} K^+$ and $p \to \mu^+ K^\circ$ modes by a factor of 2 to at most 10 should either turn up events, or else the remarkably successful $SO(10)$ framework described here will be called into question seriously. On the basis of our study, we expect that the SuperK detector should in fact see a few proton decay events in the $\bar{\nu} K^+$ and quite possibly in the $\mu^+ K^\circ$ channel in the near future. To establish the reality of this important process firmly and also to study efficiently the branching ratios of some crucial modes, like the $\mu^+ K^\circ$, next generation detectors with sensitivity of at least 5×10^{34} and perhaps 10^{35} years are essential.

We have stressed that observation of proton decay into $\mu^+ K^\circ$ with a branching ratio exceeding 20% (say) would provide a clear signature in favor of (a) supersymmetric unification based on symmetry structures such as a string-derived $G(224)$ or $SO(10)$, as well as (b) the mechanism described here of generating the masses and mixings of all fermions including especially the neutrinos [18].

To conclude, proton decay has been anticipated for quite some time as a hallmark of grand unification. With coupling unification and neutrino masses revealed, proton decay is the missing link. While its discovery, with dominance of the $\bar{\nu} K^+$ mode, would confirm supersymmetric unification, prominence of the $\mu^+ K^\circ$ mode establish the beautiful link that exists between the neutrino masses and proton decay within the $G(224)/SO(10)$-route to unification.

Acknowledgements

I would like to thank specially Kaladi S. Babu and Frank Wilczek for a most enjoyable collaboration on the research described here. I would also like to thank Dr. Gautam Sidharth for the most kind hospitality extended to me during the symposium. The research presented here is supported in part by DOE grant NO.DE-FG02-96ER-41015.

References

1. Y. Fukuda, et al., *Phys. Lett.* **81**, 1562 (1998).
2. J.C. Pati, *Neutrino 98*, Takayama, Japan, June, 98, hep-ph/9807315; *Nuclear Phys. B* (Proc. Suppl.), **77**, 299 (1999).

3. S. Weinberg, *Phys. Rev. Lett.* **43**, 1566 (1979); E. Akhmedov, Z. Berezhiani and G. Senjanovic, *Phys. Rev. Lett.* **69**, 3013 (1992).

4. J.N. Bachall, P. Krastev and A. Yu Smirnov, *Phys. Rev. D.* **58**, 096016 (1998).

5. J.C. Pati and Abdus Salam, *Proc. 15th High Energy Conference*, Batavia **2**, 301 (1972); *Phys. Rev.* **8**, 1240 (1973).

6. J.C. Pati and Abdus Salam *Phys. Rev. Lett.* **31**, 661 (1973); *Phys. Rev. D* **10**, 275 (1974).

7. H. Georgi and S.L. Glashow, *Phys. Rev. Lett.* 32, 438 (1974).

8. H. Georgi, H. Quinn and S. Weinberg, *Phys. Rev. Lett.* **33**, 451 (1974).

9. M. Green and J.H. Schwarz, *Phys. Lett.* B **149**, 117 (1984); D.J. Gross, J.A. Harvey, E. Martinec and R. Rohm, *Phys. Rev. Lett.* **54**, 502 (1985); P. Candelas, G.T. Horowitz, A. Strominger and E. Witten, *Nucl. Phys.* B **258**, 46 (1985); M.B. Green, J.H. Schwarz and E. Witten, *Superstring Theory* Vols. 1 and 2, Cambridge University Press; ed., M. Dine, *String Theory in Four dimensions*, North Holland (1988); J. Polchinski, *Les Houches Lectures*, hep-th/9411028 (1994).

10. P. Langacker and M. Luo, *Phys. Rev. D* **44**, 817 (1991); U. Amaldi, W. de Boer and H. Furstenau, *Phys. Lett.* B **260**, (1991); J. Ellis, S. Kelley and D.V. Nanopoulos, *Phys. Lett.* B **260**, 131 (1991); F. Anselmo, L. Cifarelli, A. Peterman and A. Zichichi, *Nuovo Cim.* A **104**, 1817 (1991); S. Dimopoulos, S. Raby and F. Wilczek, *Phys. Rev. D* **24**, 1681 (1981); W. Marciano and G. Senjanovic, *Phys. Rev. D* **25**, 3092 (1982); M. Einhorn and D.R.T. Jones, *Nucl. Phys.* B **196**, 475 (1982).

11. S. Weinberg, *Phys. Rev. D* **26**, 287 (1982); N. Sakai and T. Yanagida, *Nucl. Phys.* B **197**, 533 (1982).

12. Y.A. Gelfand and E.S. Likhtman, *JETP Lett.* **13**, 323 (1971); J. Wess and B. Zumino, *Nucl. Phys.* B **70**, 139 (1974); D. Volkov and V.P. Akulov, *JETP Lett.* **16**, 438 (1972).

13. K. Dienes, *Phys. Reports* **287**, 447 (1997). hep-th/9602045, and references therein.

14. J.C. Pati, hep-ph/9811442; *Proc. Salam Memorial Meeting*, World Scientific, (1998).

15. H. Georgi, in *Particles and Fields*, ed., C. Carlson, AIP, NY, (1975); H. Fritzsch and P. Minkowski, *Ann. Phys.* **93**, 193 (1975).

16. I. antoniadis, G. Leontaris and J. Rizos, *Phys. Lett.* B **245**, 161 (1990); G. Leontaris, *Phys. Lett.* B **372**, 212 (1996).

17. K.S. Babu, J.C. Pati and F. Wilczek, *Phys. Lett.* B **434**, 337 (1998).

18. K.S. Babu, J.C. Pati and F. Wilczek, hep-ph/981538V3; *Nucl. Phys.* B, to appear.

19. S. Mikheyev and A. Smirnov, *Nuovo Cim.* C **9**, 17 (1986); L. Wolfenstein, *Phys. Rev. D* **17**, 2369 (1978).

20. J.C. Pati and A. Salam, *Phys. Rev. D* **10**, 275 (1974); R.N. Mohapatri and J.C. Pati, *Phys. Rev. D* **11**, 566, 2558 (1975); G. Senjanovic and R.N. Mohapatra, *Phys. Rev. D* **12**, 1502 (1975).

21. V. Kuzmin, Va Rubakov and M. Shaposhnikov, *Phys. Lett.* BM **155**, 36 (1985); M. Fukugita and T. Yanagida, *Phys. Lett.* B **174**, 45 (1986); M.A. Luty, *Phys. Rev. D* **45**, 455 (1992); W. Buchmuller and M. Plumacher, hep-ph/9608308.

22. F. Gürsey, P. Ramond and P. Sikivie, *Phys. Lett.* B **60**, 177 (1976).

23. P. Langacker and N. Polonsky, *Phys. Rev. D* **47**, 4028 (1993) and references therein.

24. E. Witten, *Nucl. Phys.* B **443**, 85 (1995); P. Horava and E. Witten, *Nucl. Phys.* B **460**, 506 (1996); J. Polchinski, hep-th/9511157; A. Sen, hep-th/9802051, and references therein; M. Duff, hep-ph/9805177 V3.

25. P. Ginsparg, *Phys. Lett.* B **197**, 139 (1987); V.S. Kaplunovsky, *Nucl. Phys.* B **307**, 145 (1988); Erratum, ibid. **382**, 436 (1992).

26. E. Witten, hep-th/9602070.

27. J.C. Pati and K.S. Babu, hep-ph/9606215, *Phys. Lett.* B **384**, 140 (1996).

28. D. Lewellen, *Nucl. Phys.* B **337**, 61 (1990); A. Font, L. Ibanez and F. Quevedo, *Nucl. Phys.* B **345**, 389 (1990); S. Chaudhari, G. Hockney and J. Lykken, *Nucl. Phys.* B **456**, 89 (1995) and hep-th/9510241; G. Aldazabad, A. Font, L. Ibanez and A. Uranga, *Nucl. Phys.* B **452**, 3 (1995); ibid. B **465**, 34 (1996); D. Finnell, *Phys. Rev.* D **53**, 5781 (1996); A.A. Maslikov, I. Naumov and G.G. Volkov, *Int. J. Mod. Phys.* A **11**, 1117 (1996); J. Erler, hep-th/9602032 and G. Cleaver, hep-th/9604183; and Z. Kakushadze and S.H. Tye, hep-th/9605221, and hep-th/9609027, Z. Kakushadze et al., hep-ph/9705202.

29. A. Faraggi, *Phys. Lett.* B **278**, 131 (1992); *Phys. Lett.* B **274**, 47 (1992); *Nucl. Phys.* B **403**, 101 (1993); A. Faraggi and E. Halyo, *Nucl. Phys.* B **416**, 63 (1994).

30. See e.g. [16]

31. J.C. Pati, hep-ph/9607446, *Phys. Lett.* B **388**, 532 (1996).

32. A. Faraggi and J.C. Pati, hep-ph/9712516v3, December (1997), *Nucl. Phys.* B (to appear).

33. A. Faraggi and J.C. Pati, *Phys. Lett.* B **400**, 314 (1997).

34. K.S. Babu and J.C. Pati, *Towards a resolution of the supersymmetric CP problem through flavor and left-right symmetries*, (to appear).

35. K.R. Dienes and J. March-Russell, hep-th/9604112; K.R. Dienes, hep-ph/9606467.

36. M. Gell-Mann, P. Ramond and R. Slansky, in *Supergravity*, (eds.) F. van Nieuwenhuizen and D. Freedman, Amsterdam, North Holland (1979) p. 315; T. Yanagida, in: *Workshop on the Unified Theory and Baryon Number in the Universe* (eds.), O. Sawada and A. Sugamoto KEK, Tsukuba, 95 (1979); R.N. Mohapatra and G. Senjanovic, *Phys. Rev. Lett.* **44**, 912 (1980).

37. H. Georgi and C. Jarlskog, *Phys. Lett.* B **86**, 297 (1979).

38. F. Wilczek and Z. Zee, *Phys. Lett.* B **70**, 418 (1977); H. Fritzsch, *Phys. Lett.* B **70**, 436 (1977).

39. C. Albright, K.S. Babu and S.M. Barr, *Phys. Rev. Lett.* **81**, 1167 (1998).

40. J. Gasser and H. Leutwyler, *Phys. Rept.* **87**, 77 (1982).

41. R. Gupta and T. Bhattacharya, *Nucl. Phys. Proc. Suppl.* **53**, 292 (1997); and *Nucl. Phys. Proc. Suppl.* **63**, 45 (1998).

42. V. Barger, M. Berger and P. Ohmann, *Phys. Rev.* D **47**, 1093 (1993); M. Carena, S. Pokorski and C. Wagner *Nucl. Phys.* B **406**, 59 (1993); P. Langacker and N. Polonsky, *Phys. Rev.* D **49**, 1454 (1994); D.M. Pierce, J.A. Bagger, K. Matchev and R. Zhang, *Nucl. Phys.* B **491**, 3 (1997); K.S. Babu and C. Kolda, hep-ph/9811308.

43. S. Dimopoulos, S. Raby and F. Wilczek, *Phys. Lett.* B **112**, 133 (1982); J. Ellis, D.V. Nanopoulos and S. Rudaz, *Nucl. Phys.* B **202**, 43 (1982).

44. P. Nath, A.H. Chemseddine and R. Arnowitt, *Phys. Rev.* D **32**, 2348 (1985); P. Nath and R. Arnowitt, hep-ph/9708469.

45. J. Hisano, H. Murayama and T. Yanagida, *Nucl. Phys.* B **402**, 46 (1993).

46. K.S. Babu and S.M. Barr, *Phys. Rev.* D **50**, 3529 (1994); D **51**, 2463 (1995).
47. Y. Hayato, *Int. Conf. High Energy Physics*, Vancouver, July (1998).
48. See e.g. [45].
49. S. Dimopoulos and F. Wilczek, Report No. NSF-ITP-82-07 (1981), *Proc. of the 19th Course of the International School on Subnuclear Physics* (ed.), A. Zichichi), Erice, Italy, 1981, Plenum Press, New York; K.S. Babu and S.M. Barr, *Phys. Rev.* D **48**, 5354 (1993).
50. N. Tatsui et al., *JLQCD collaboration*, hep-lat/9809151.

The Nature of Discovery in Physics

Douglas D. Osheroff

Stanford University, California, U.S.A

Fig. 1. Douglas D. Osheroff delivering the B.M. Birla Memorial Lecture

B.G. Sidharth (ed.), *A Century of Ideas*, 175–203.

Douglas (Doug) Dean Osheroff was born in Aberdeen, Washington State, the second of five children to parents in a very noble profession. His father was a physician while his mother was a nurse. The young Doug had a more than usual "hands on" childhood. He ripped toys and even gadgets apart under the tolerant eye of his parents. Occasionally his adventure with gadgets turned out to have potentially dangerous consequences.

Subsequently he had the good fortune to study Physics at the California Institute of Technology, where, he was a student of Richard Feynman, amongst others. It was here that he met a student from Taiwan, Phyllis, whom he married.

In 1971, working at Caltech, he discovered the superfluidity of Helium-3. He then moved on to the east coast, where he received a PhD from Cornell University in 1973. Between 1972 and 1987 Doug was on the technical staff of AT&T Bell Laboratories where he became the Head of the Low Temperature and Solid State Research Department. In 1987 he moved to California, joining the Stanford University where he has been a Professor of Physics and Applied Physics through all these years. He has continued his studies on superfluid and solid Helium-3.

Prof. Osheroff received the Nobel Prize for Physics for his earlier work on the superfluidity of Helium in 1996. This apart he has received numerous other distinctions and awards. These include the Francis Simon Memorial Award, Oliver E Buckley Condense Matter Physics Prize, Mac Arthur Prize Fellowship Award, the Walter J. Gores Award for excellence in teaching and so on. He was also a member of the Columbia accident investigation board.

Prof. Osheroff is a very enthusiastic and lively physicist with a soft human side. Recently he wrote: "I will only indicate that I find that human population and our ability to consume raw materials and to produce wastes has left us with a planet unable to cope with our consumption. I think the most pressing problem for mankind at present is to produce a renewable economy, and stop polluting our environment. Global warming is probably the most pressing problem, particularly with the development going on in China and India now. If our production of greenhouse gases leads to the melting of the polar ice caps, (including Greenland), we will be in very deep trouble."

In his B.M. Birla Memorial Lecture Prof. Osheroff delightfully recounts his childhood and research experiences. The ensuing article is reprinted with permission from the American Journal of Physics, Volume 69, pages 26–37, 2001, American Association of Physics Teachers.

Summary. By their very nature, those discoveries which most change the way we perceive our physical universe are difficult to anticipate. How then, are such discoveries made, and what experimental approaches are most likely to lead to discoveries? In this article I will describe four experiments in which I have participated that have yielded unexpected new physics, and attempt to explain how they came about.

1 Introduction

It is often said that to make an important discovery in physics one must either be good or be lucky, but that good people manufacture their own luck. That is to say, some people have a knack for finding new things, either because they manage to be at the right places at the right times, or because they are able to recognize subtle evidence that physics is different than what people had believed. To be at the right place at the right time is not always, if ever, a matter of pure luck. Frequently the people involved have recognized the potential for new physics when a new technology has allowed scientists to study nature in a new realm, or allowed them to observe some aspect of our physical universe more clearly than ever before. I will first illustrate these ideas by describing the nature of two discoveries, which have played an important role in shaping the course of physical research even to this very day, and are subtly connected to one another.

The first discovery I wish to consider is that of superconductivity, made by Kammerlingh-Onnes in 1911 [1]. This discovery would have been virtually impossible were it not for the new technology developed by Kammerlingh-Onnes which allowed him to liquefy helium for the first time in 1908. In so doing, he won the titanic struggle with Dewar to liquefy the lightest of the atmospheric gases. Once Kammerlingh-Onnes found that he could not solidify helium under its own vapor pressure, even at temperatures as low as 1.04 Kelvin, he turned his attention to the question of what happened to electrical conduction in metals at very low temperatures. One speculation was that the conduction electrons would 're-condense' onto their parent atoms and thus the electrical resistance would rise toward infinity as T approached zero. Another possibility was that as T approached zero in a very pure metal, the lattice vibrations would slowly disappear, and thus the resistance would decrease toward zero.

In 1911 Kammerlingh-Onnes directed his student, Giles Holst, with the help of his technician G.J. Flim, to measure the resistance of a sample of very pure mercury. What those two found was neither of the above. Instead, they found that the resistance dropped nearly to zero over a very narrow range in temperature, far more rapidly than expected. While this observation did not raise too many eyebrows when first presented in 1911, within two years it was clear that this 'super-conductive state' could also be produced in lead and tin, but not in such metals as gold and platinum. Heike Kammerlingh-Onnes was awarded the Nobel Prize for Physics in 1913 "for his investigations on the properties of matter at low temperatures which led, inter alia, to the production of liquid helium." In his Nobel Prize lecture he highlighted this remarkable superconductive state. While superconductivity has been studied extensively from 1911 to the present, its microscopic origins remained a mystery until Bardeen, Cooper, and Schrieffer produced their BCS theory of superconductivity in 1957 [2]. Ironically, those three were awarded the Nobel Prize for physics in 1972, the same year that superfluidity in ^{3}He was discovered, a phenomenon which is described by very similar physics.

In the course of his studies of superconductivity, Kammerlingh-Onnes failed to ever recognize that the helium bath in his cryostat itself underwent a transition to a superfluid state, even though he most certainly created this remarkable substance numerous times. It is perhaps worth considering why Kammerlingh-Onnes was able to discover superconductivity but not superfluidity. It is well known that the thermal conductivity of liquid helium becomes almost infinite just below the superfluid transition at $2.17\,K$, almost as dramatic and abrupt a change in behavior as is exhibited by the electrical conductivity at the superconducting transition. The reason he succeeded in one case and not the other had to do with the fact that current speculation led him to study the conductivity of metals, but that there was nothing which suggested that the thermal conductivity of liquid helium should do anything interesting as T approached zero.

It was not until the mid-1930s that the remarkable nature of this superfluid became clear. While no one received a Nobel prize for the discovery of superfluidity in liquid helium, Peter Kapitza was granted half the Nobel Prize for Physics in 1978 "for his basic inventions and discoveries in the area of low temperature physics."

The other half of the Nobel Prize for physics in 1978 was shared by Arno Penzias and Robert Wilson, for their discovery of the cosmic microwave background radiation [3]. This discovery is the other example I wish to discuss, and perhaps the most important contribution to astrophysics in the past half century. This is another interesting discovery, for it was not something which Penzias and Wilson had set out to find, nor was it obvious to them what they had discovered. The two, working at the Bell Laboratories facility at Crawford Hill, had just built the world's most sensitive instrument for making absolute measurements of the sky brightness at various wavelengths. To do this, they switched their microwave receiver from a horn antenna to a reference source as liquid helium temperatures. They had set about to measure the background radiation from the halo of our galaxy, and pointed their antenna at a 'quiet' part of the sky. Much to their surprise, the sky brightness they found had a characteristic noise temperature of $3\,K$, much higher than expected. Extensive repairs to their horn antenna did not significantly improve the situation, but the two could not dismiss the elevated noise level in their receiver.

Ultimately, Penzias heard from a colleague at MIT that Robert Dicke, a professor at Princeton University, had postulated the existence of fossil radiation left over from a primordial explosion in a model of an oscillating universe (Dicke was not, however, the first to have done so [4]). Penzias phoned Dicke, who promised to send him a preprint of a theory paper by P.J.E. Peebles in his group. Dicke and his team, including instructors David Wilkinson and Peter Roll, were in the process of searching for this fossil radiation. As Dicke hung up the phone after the conversation with Penzias, he announced to his group: "Gentlemen, we've been scooped [5]." Soon the entire Princeton group visited Penzias and Wilson at Crawford Hill. Ultimately the two groups agreed to publish back to back theory and experimental papers in Astrophysical

Journal. Had Penzias and Wilson brushed aside the apparent excess noise in their receiver, it seems likely that Dicke would have been the discoverer of the cosmic microwave background radiation.

Penzias and Wilson were first largely because they had the best technology. The travelling wave tube maser amplifier which Penzias and Wilson used had a noise temperature of $20\,K$, while the microwave tube used by Dicke's group had a noise temperature of $2500\,K$ [5]. It was Penzias and Wilson's good fortune to be at Bell Laboratories, which was interested in creating the most sensitive microwave receivers for use in satellite communications experiments. Clearly technology played an important role in this discovery. Penzias and Wilson would not have made the discovery, however, had they been less concerned about the performance of their receiver, or less confident in their understanding of its limitations.

Did Dicke contribute to the discovery, or did he simply help explain something which had already been discovered? Nobel prizes are frequently given for important discoveries, and in those instances, it is typically only the discovery, and not the elucidation of the new phenomena which is rewarded. I like to think of such prizes as recognizing the importance of new physics, and celebrating its discovery. Some of the fame and credit which accompanies most Nobel prizes should go to others, but does not because the media, the public, and even the scientific community like to over-simplify history, and seldom bother to understand how these discoveries actually took place.

In the remainder of this article, rather than discussing other people's great discoveries, I shall describe four discoveries in which I have participated. Not all of these discoveries are very important, and only one of them has been recognized by a prize of any sort. They all serve, however, to illustrate the following ideas: (a) The period of discovery in physics is often a chaotic one, in that those making the discovery seldom recognize immediately the nature of what they have found. (b) Certain strategies in experimental work can greatly enhance the probability of discovery without significantly decreasing one's productivity in the event that there is no new physics to be found. Before I begin that discussion, however, I will include a brief account of my personal background, for it may help the reader to understand my perspective on scientific research.

2 Personal Background

I grew up Aberdeen, Washington, a logging town in the Pacific Northwest, just below the Olympic Peninsula. I was the second of five children. When I was young, my brothers and I frequently took extended walks along long deserted logging roads that wound their way deep into the hemlock and fir forests surrounding Aberdeen. We would imagine ourselves as explorers, perhaps the first to have ventured so deep into these (second growth) forests. To me, life was an adventure, and discovery was the ultimate reward, be it a long deserted cabin, or a rusty iron ball which I imagined had been shot from some sailing vessel which long ago had visited the shores of Grays Harbor.

As a young child I had a natural curiosity about how things worked, and my parents never discouraged this curiosity. For example, I tore my first electric train engine apart at age six to play with the electric motor. At about age eight, my father gave me the camera he had used as a child, and within hours it lay in tiny pieces all over the living room carpet, never to be reassembled. My father did not scold me for this, but helped to satisfy and direct my natural curiosity. Our house was soon filled with all manner of scientific toys. Erector sets, Tinker toys, Gilbert chemistry sets, American Optics sets, electronics sets, and eventually Heath Kit radios. But more than anything else, I enjoyed taking things apart and putting them back together.

My father was a physician, and often his patients would give him things for me to investigate. Shortly after the incident with the camera, he brought me a used watch and a set of jewellers screw drivers. I soon learned how to take watches apart and put them back together. They would always work, although I invariably had parts left over. When I was still in grade school, my father brought me some surplus parts from the local Bell Telephone switching office. I found that if I connected the leads from the solenoid of a relay to a $22\,V$ battery, I would get a shock when I removed the leads. I didn't understand inductance, but I packaged the device in a gym bag so that the circuit would be closed when someone squeezed the prongs of lamp plug together. They would then get a mild shock when they let the prongs move apart. At recess my classmates lined up to get shocked by my 'machine'.

As I entered my teenage years, I discovered chemistry, gunpowder, and high voltage electricity. This was perhaps the most dangerous period of my life. I was fascinated by gunpowder, and soon rockets, bombs, and cannons occupied my spare time. Once a muzzle loading rifle I had built went off in the house, putting a hole through two walls. My parents and I had an understanding. When I did something really dumb, I would simply stop that activity, and they wouldn't bother lecturing me. Perhaps the closest I came to serious injury was the time I had built a calcium carbide 'miner's lamp' but lit the flame before all the oxygen had escaped from the beaker I used to contain the calcium carbide and water. Instead of getting a brilliant white flame, I got a pale blue one. I instinctively turned my head away from the apparatus just before it exploded. There was glass sticking in the walls, and in the side of my face. My mother was upstairs preparing dinner, and at the sound of the explosion she came to the top of the stairs. I was coming up the stairs, cupping my hands so as to keep the blood from dripping onto the carpet. Knowing fully my propensity for practical jokes, she shouted out "If you're kidding I'll kill you." As my father sewed the worst of my cuts closed he lectured me about safety. In another instance I managed to discharge a bank of capacitors charged to $600\,V$ across my body. The resulting muscle contraction propelled me across the room, and the next thing I knew I was lying on the floor against the far wall.

This 'experimentation' continued until I went off to college. While it was only tinkering and in no way true science, I acquired very good technical skills, and developed excellent physical intuition. My first introduction to scientific observation came in high school chemistry class. My teacher, William Hock, had been involved in research, and wanted to give his class a flavor of what research was like. We spent time making our own observations of a burning candle, and when I simply wrote down how I understood a candle to function, he asked me to re-do the exercise. Mr. Hock one day brought a milk carton to class with what I believe was a cloths pin inside. He likened research to the process which we might use to try to decide what was inside. I wished that I had had such an experience earlier in my schooling.

I went to Caltech for my undergraduate education. I arrived while Richard Feynman was still teaching his famous undergraduate course, and physics was clearly king among undergraduates at Caltech. I managed to graduate with honors, but was not near the top of my class. In my junior year I began to loose focus, and found many excuses for not studying. I was wondering if I had a future in physics, which appeared to be nothing but endless problem sets. Fortunately, a friend named Andy McKay asked me if I was interested in working in a research group, for money. I don't know what was more important to me, the activity, or the money, but I started immediately working in the infrared astrophysics group directed by Gerry Neugebauer. Infrared astrophysics was a relatively new field, and there were many exciting things happening. Although my first job only involved correcting parity errors in punched paper tape, my last was to help design and build an IR camera for creating some of the first images of the center of the Milky Way galaxy.

While I enjoyed the work in astrophysics, I soon realized that astrophysicists seldom actually perform experiments. They mostly observe distant objects. I wanted to be more than an observer in the science I studied, I wanted to participate. I wanted to control the systems I studied, to poke them, to prod them, and to make them tell me their deepest secrets. Fortunately, I became involved in David Goodstein's low temperature program during my senior year. I suppose that involvement ultimately had a profound impact on my career, even though my own work in the lab did not actually involve low temperatures.

I decided by the spring of my senior year that I wanted to go into condensed matter physics, but choose Cornell for graduate school largely out of ignorance. David Goodstein was at that time in Italy, and there was no one in the Physics Department at Caltech who knew much about condensed matter physics (I didn't dare ask Richard Feynman for such trivial advice). Fortuitously, it was a wonderful time to be at Cornell, and my decision was a lucky one, for a little over four years after arriving at Cornell I was to make the first of my four discoveries.

3 The Discovery of Superfluidity in Helium Three

My first year at Cornell I supported myself on a teaching assistantship, as was the custom for students without fellowships. I taught for David Lee, who had built up the low temperature group at Cornell. Dave thought that Caltech students were smart, and encouraged me to visit his lab. I did so, and found there to be a lot going on. I was attracted to low temperature physics because of new and powerful cooling technologies being developed, which I felt held promise to allow physicists to look at nature in a realm in which she had never been seen before. One of those technologies was the ^{3}He-^{4}He dilution refrigerator, capable of maintaining temperatures as low as $15\,mK$ for extended periods of time. My second semester at Cornell I began building the dilution refrigerator which would ultimately be used in my first discovery.

The most impressive of the new technologies was based on a conjecture made by I. Pomeranchuk in 1950 [6]. At this time little was actually known about the low temperature properties of liquid or solid ^{3}He. Pomeranchuk, however, recognized that since ^{3}He atoms had a net spin of 1/2, they must be Fermi particles, as are electrons. Thus, at low temperatures, the liquid would possess an entropy (the degree of disorder) and a heat capacity linear in the temperature, just as one finds for conduction electrons in metals. In the solid, by contrast, the nuclear spins would be oriented randomly, providing an entropy of nearly Rln(2) which would be nearly independent of temperature. Ultimately, the nuclear spins in the solid should order either ferromagnetically or antiferromagnetically, but Pomeranchuk didn't expect that would happen at temperatures above a microkelvin.

Thus, below some temperature (which we now know is $314\,mK$), the disorder in the solid would exceed the disorder in the liquid, and the latent heat of solidification would become negative. Thus, to solidify the liquid at constant temperature, one would have to add heat to the system. If one performed the compression necessary to form solid adiabatically and reversibly, the system would cool. Pomeranchuk felt that this cooling process would allow one to reach temperatures well below a thousandth of a degree. The entropies of solid and liquid ^{3}He as a function of temperature are shown in Fig. 2.

Within a few thousandths of a degree of absolute zero the melting pressure of ^{3}He is about 34 Bars, and the work done in forming the solid, $p\Delta V$, would exceed the cooling capacity of the process, $T(S_{solid} - S_{liquid})$ by as much as three orders of magnitude. If any of the work expended in forming solid showed up as heat, the process might actually lead to heating rather than cooling. Thus no one tested Pomeranchuk's idea until 1965, when a student named Anufriyev in Kapitza's lab found that he could cool a sample of liquid from about $60\,mK$ to slightly below $20\,mK$ by Pomeranchuk refrigeration [7]. While this was not as low a temperature as could be reached continuously using a ^{3}He-^{4}He dilution refrigerator, it seemed plausible that Pomeranchuk was correct, and thus the process held the promise of allowing low temperature researchers to study ^{3}He in a new realm. Dave Lee was one of a few physicists in the United States who decided to develop the Pomeranchuk process further.

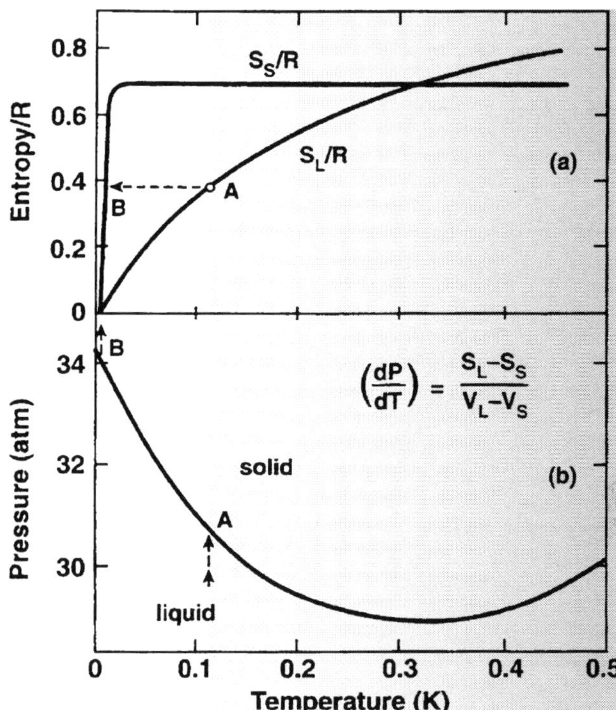

Fig. 2. (a) The entropies of liquid and solid ^{3}He along the melting line as a function of temperature cross at about $314\,mK$. Below this temperature the liquid is more highly ordered than the solid, and the latent heat of solidification is negative. **(b)** The slope of the coexistence line between the liquid and solid phases is given by the Clausius-Clapeyron equation. Below the point where the solid and liquid entropies are equal, the slope must be negative. Pomeranchuk argued that if one compresses the liquid, one will move along the melting line from A to B as the fraction of the sample in the solid phase goes from 0 to 1

By this time it was recognized that spin ordering in solid ^{3}He at melting pressure would probably occur at a temperature of about $2\,mK$, much higher than Pomeranchuk had expected. This would result from the rapid exchange of ^{3}He atoms between neighboring lattice sites, at frequencies as high as 4.10^7 times per second. This didn't bother Dave, however, because it was precisely this ordering which he intended to study. I worked hard to put myself in charge of the cryostat in which this study would be carried out.

It was not superfluidity in ^{3}He which we were after when we began the experiment which was to lead to its discovery. Indeed, by the time I became a graduate student at Cornell, most people doubted that such a state of matter would ever be found. Hopes had been high [8] shortly after the publication of the BCS theory which explained the origins of superconductivity in 1957 [2]. It was considered likely that a similar mechanism would lead to superfluidity in liquid ^{3}He as well. Initial estimates of the transition temperature, T_c, were overly optimistic, as high as $100\,mK$. However, when experimental searches found nothing, the theorists were able to generate lower estimates, leading to a new round of more difficult experiments. Eventually the experimentalists had cooled the liquid down to about $2\,mK$, and found nothing. At this point the theorists decided they had chosen the wrong Cooper pair interaction (needed to correlate the Fermionic ^{3}He atoms together into Bosonic entities), and finally predicted a T_c of about $50\,\mu K$. At this point all experimental effort ceased, for no one knew how to reach this very low temperature.

Near the end of my fourth year of graduate study, Dave Lee suggested that I use Pomeranchuk refrigeration to measure some property of solid He^3 through the nuclear spin ordering transition for my PhD. thesis. However, at this time no one was able to create a thermometer which would stay in good thermal contact with the liquid ^{3}He below about $3\,mK$. I was stuck, until Dave handed me a preprint from the Wheatley group at UCSD [9]. They had studied how a strong magnetic field affected the ^{3}He melting pressure, and what they found was most surprising. Low magnetic field seemed to suppress the melting pressure by an amount which was more than a factor of ten greater than simple thermodynamic calculations suggested. This all occurred above $5\,mK$, where our thermometers were known to work well. Indeed, the Wheatley group and I had both measured the melting pressure in weak magnetic fields down to $2.8\,mK$, and our results, with very different thermometry techniques, agreed. I decided to reproduce the UCSD result.

The UCSD result was not correct, but an artifact of bad thermometry. Yet this result steered us in the right direction, just as the possibility that conduction electrons might re-condense on their parent atoms in metals had caused Kammerlingh-Onnes to look at the conductivity of metals at very low temperatures. When I tried to reproduce the dramatic results found by Wheatley and his students, I observed only the very small suppression which was predicted by thermodynamics. Because the result was so small, my measurements of it could not be very precise, and I struggled to improve my measurement technique.

This was a hopeless experiment, and I was indeed fortunate that two other students in the lab convinced Professors Dave Lee and Bob Richardson that I should relinquish the lab's only NMR quality electromagnet so that they could carry out a different experiment. I determined the temperature by using NMR to measure the polarization of copper nuclei along a weak magnetic field, and without the electromagnet I could not continue my experiment.

Things were not going very well. My experiment looked like a flop, and now I was forced to give up an essential piece of my equipment. Looking back at that time, I am a bit amazed that I didn't warm up my cryostat and take a short vacation. But instead, I decided to keep my cryostat cold until it was clear whether or not the other experiment would work. If a leak opened up or a wire fell off inside their cryostat, the other students would be forced to warm up, and I would get the magnet back.

This fortuitous sequence of events gave me a working cryostat, freed from the burden of a bad experiment. I decided to test whether or not this Pomeranchuk refrigerator could indeed reach the temperature of $2\,mK$ at which one expected the solid nuclear spin system to order. Since my thermometer wouldn't indicate the correct temperature below $2.7\,mK$ anyway, it didn't matter that it was unavailable. Instead, I felt that I could monitor the temperature by measuring the melting pressure, which was related to the temperature through the Clausius-Clapyeron equation:

$$\frac{Dp_{melt}}{dT} = \frac{s_{solid} - S_{liquid}}{V_{solid} - V_{liquid}}$$

Here S and V are the molar entropy and volume respectively. Notice that this equation shows that the melting pressure will have a minimum at the temperature for which the two entropies are equal, and that it rises with a negative slope below that temperature, as can be seen in Fig. 2b above. Assuming that the entropy of the solid did not change too rapidly below $2.7\,mK$, I felt I could extrapolate the measured melting line from $2.7\,mK$ to $2\,mK$ with reasonable certainty.

The first test of this idea occurred on November 24, 1971. This was the day before Thanksgiving. I decreased the volume of my cell at a steady rate by extending a metal bellows into the ^{3}He filled region. This caused solid ^{3}He to form at a constant rate, and the system cooled as I had seen many times before. I started at about $22\,mK$, and the rate of cooling was about $1\,mK$ for every percent of the liquid sample converted to solid. In a Pomeranchuk cell one had no control over where the solid ^{3}He formed, and my fear was that ultimately the metal bellows would begin to deform solid ^{3}He, and this would most surely lead to heating.

My fear appeared to become reality as my cell reached a temperature I estimated to be about $2.6\,mK$. There was a sharp decrease in the rate of cooling, by about a factor of three, and I terminated the experiment shortly after this, greatly disheartened. However, this was the day before Thanksgiving, and I felt that if I allowed my apparatus to pre-cool over the entire four day holiday, I could repeat the experiment the next Monday, starting at a much lower temperature. This would mean that at every temperature I would have less solid ^{3}He in the cell, and hence there would be much less likelihood that I would run into the heating problem again.

The next Monday I started my compression at about $15\,mK$. Thus at $2.6\,mK$ I should have about 30% less solid in my cell than in my first trial. As I passed through the temperature region where I had run into problems in my first compression, however, I again saw a sharp decreasein the rate of cooling. Upon closer examination, I found that the melting pressures at which the two events had occurred were the same to within one part in 50,000. It seemed extremely unlikely that such close agreement could be a coincidence, particularly given the very different starting conditions. I concluded that this sharp kink in the pressurization curve must result from some highly reproducible

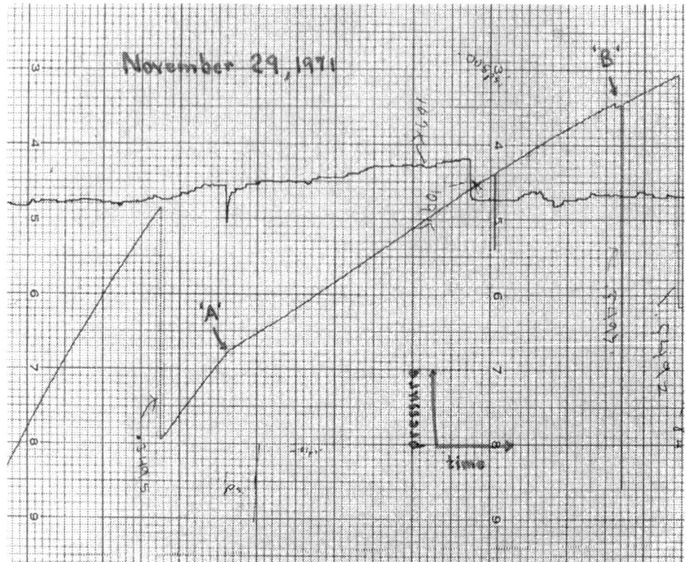

Fig. 3. The ^{3}He pressure in the Pomeranchuk cell as a function of time as liquid is converted to solid at a constant rate. The features marked A and B represent the new phase transitions discovered in this experiment

phase transition within the mixture of solid and liquid ^{3}He in my cell. But no transition was expected in either the liquid or the solid at this temperature. This second pressurization curve is shown in Fig. 3. Notice the feature labelled 'B' near the right had edge of the figure. This was later determined to be a second unexplained phase transition.

The rest of this story is chronicled in my Nobel lecture [10], and I won't repeat the details here. It is a tale of confusion, of desperate actions, and of eventual understanding after months of experimental study. We first guessed that the transitions we had discovered were occurring in the solid phase, but ultimately recognized that Dave Lee, Bob Richardson and I had discovered three new states of liquid ^{3}He, states which had unusual magnetic resonance properties, never before predicted to exist. We believed they were most probably BCS superfluid phases. A fuller understanding came over a three year period of intense world-wide study. The interested reader is encouraged to read about these efforts as well [10].

Before going on to the second discovery, let me recount the things which I believe I did right, and did wrong leading up to this discovery. (1) I believe I was attracted to low temperature physics for the right reasons: It was a field ripe for discovery because of the new cooling technologies which were being developed. Just as Kammerlingh-Onnes had done, we were able to see nature in a new realm. (2) The measurement which drew us to the discovery was a wrong speculation, this time based on experimental data. However, as

I continued to repeat my own measurements of the shift in melting pressure with magnetic field, almost as if I were a horse with blinders on, good fortune played a role, forcing me to try something new. (3) We were sufficiently confident that we understood our apparatus to recognize the potential importance when something unexpected presented itself. (4) One important point not covered in the above discussion has to do with outside commitments, particularly made by graduate students. In August, 1970 I married my present wife of 29 years, who grew up in Taiwan. In the fall of 1971 Cornell was offering a new course on conversational Chinese, and my wife and I had agreed that I would take that course. Eventually I realized that this was my last year of graduate study, however, and that I didn't need any distractions from my thesis work. Had I taken that course, I question whether I would have kept my cryostat cold and done the experiment which ultimately lead to a Nobel Prize. I believe everyone, including graduate students, needs relaxation and occasional diversion, but I strongly recommend diversions which do not force one to adhere to a fixed schedule. The demands of research should dictate the calendar of a graduate student.

I left Cornell in the summer of 1972 for a regular position in the research area at *AT&T* Bell Laboratories. Bell Labs was very supportive of my new mission in life, to understand the newly discovered phases of liquid ^{3}He. I was able to work with some of the best theorists in the country, and was given all the experimental apparatus I could possibly use. It was probably the most exciting time of my life, and I worked as a man possessed to prove to myself and to the world that I wasn't just a lucky graduate student who managed to be at the right place at the right time. This demanding but very straightforward lifestyle continued for five wonderful years.

Finally, one day I was called into the office of the director of the physical research laboratory, Joe Burton. Joe pointed out that over the past five years I had done nothing but study superfluidity in ^{3}He. Wasn't it time that I do something else? I was not happy to hear this remark, for I had spent the past five years operating under the assumption that superfluidity in ^{3}He was the most fascinating substance one could possibly study. However, I was in industry, not academia, and the people who paid my salary also supported my research. I soon began to welcome Burton's remark, for it freed me from doing good but decreasingly exciting incremental research, and allowed me to search the physics landscape for something even more interesting. I believe that the next two years were the most productive of my physics career.

By this time few people were still using Pomeranchuk refrigeration. Adiabatic demagnetization of copper nuclei had become the dominant refrigeration technique, and was capable of cooling liquid ^{3}He at melting pressures to below $0.3\,mK$, and could keep it below $1\,mK$ for many weeks. I thought hard about what new areas of research made sense for me, given my talents, my temperament, and my experimental equipment. Ultimately two things came to mind. The first was to go back and finish my PhD. thesis work, that is, to more fully understand the nature of nuclear spin ordering in solid ^{3}He. The second was to study the breakdown of metallic conductivity in one dimensional wires. I embarked on both studies simultaneously, but will discuss the work on solid ^{3}He first.

4 Nuclear Spin Ordering in Solid Helium Three

The history of the understanding of nuclear spin ordering in solid ^{3}He is an interesting one, which I recount here only very briefly. The interested reader is referred to a review article which I wrote for a summer school in 1992, and references contained therein [11].

The nuclear dipole–dipole energy between ^{3}He nearest neighbors in the solid is equivalent to a thermal energy, $k_B T$, with T 0.1 μK. Yet if one measures the heat capacity of the solid near melting pressure, one finds that it is dominated by the changing nuclear spin entropy even at a temperature as high as $150\,mK$. Further, as the solid density increased, the apparent contribution from the nuclear spins decreased rapidly, in the opposite direction to what one would expect if direct dipole–dipole interactions were responsible for the increasing order in the solid spin system with decreasing temperature. As surprising as this behavior might seem, it was actually anticipated by Bernardes and Primakoff [12] in 1960, in the absence of any experimental knowledge of the low temperature properties of the solid (although these authors overestimated the spin ordering temperature by two orders of magnitude).

The effective spin–spin interaction used by Bernardes and Primakoff in their calculations results from the actual exchange of ^{3}He atoms between nearby lattice sites. The importance of this process was soon established, largely through NMR measurements. As the density of the solid increased, the ability of the the atoms to change places decreased because of the interference caused by the other nearby atoms.

It was measurements of the nuclear magnetic susceptibility of the solid as a function of temperature which showed most directly that the effective interactions between ^{3}He atoms would lead to antiferromagnetic order. As a second year graduate student, I was involved in one such investigation [13], whose data are shown in Fig. 4 below. Here $1/\chi$ is plotted vs. temperature, and the extrapolation to $1/\chi = 0$ shows a negative temperature, suggesting antiferromagnetism. All of the available measurements appeared to support the expectation that the nuclear spin system would undergo a second order antiferromagnetic phase transition with a transition temperature of about $2\,mK$ at melting pressures. It was this transition which I had intended to investigate in my thesis work, but never got back to after the discovery of superfluidity in liquid ^{3}He.

My thesis work had shown that the above simple model of spin–spin interactions was not adequate to describe low density solid ^{3}He. The actual transition temperature had to be at a considerably lower temperature. Measurements of the solid nuclear magnetism that I had obtained just above that

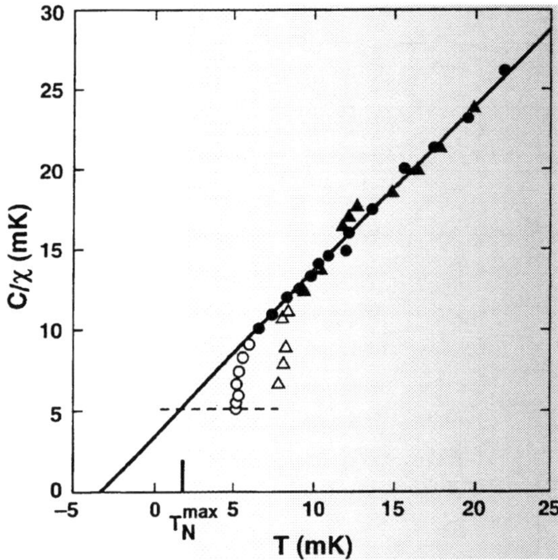

Fig. 4. The magnetic temperature, T^*, which is the Curie constant divided by the magnetic susceptibility of solid ^{3}He, is plotted as a function of temperature in the experiment of Sites, Osheroff, Richardson and Lee. The negative intercept of $T^* vs\ T$ at about $-3\,mK$ suggests that the solid should order antiferromagnetically at about $2\,mK$ at melting pressures

solid ordering had been interpreted as evidence for a competition between ferromagnetic and antiferromagnetic spin–spin interactions. This was possible if the atom–atom exchange did not occur just between nearest neighbor pairs as had been expected, but in rings of three and even four atoms. It can be shown that exchange involving an even number of atoms produces antiferromagnetic interactions, but exchange between an odd number of atoms leads to ferromagnetic interactions.

It was Bill Halperin at Cornell University, one of Bob Richardson's graduate students, who succeeded in first measuring a property (the entropy) of solid ^{3}He through the nuclear spin ordering transition [14], which he found to be at about $1\,mK$, rather than $2\,mK$. In addition, he found that the transition was strongly first order, with a large fraction of the spin entropy disappearing at T_c.

Halperin had devised a thermodynamic mechanism to measure the entropy and temperature of the ^{3}He without recourse to an external thermometer. He would apply a heat pulse to his sample, and then find out how much solid he had to form to cool back to the initial temperature. This rather clever technique was then borrowed by Dwight Adams at the University of Florida, who traced the transition temperature as a function of applied magnetic field.

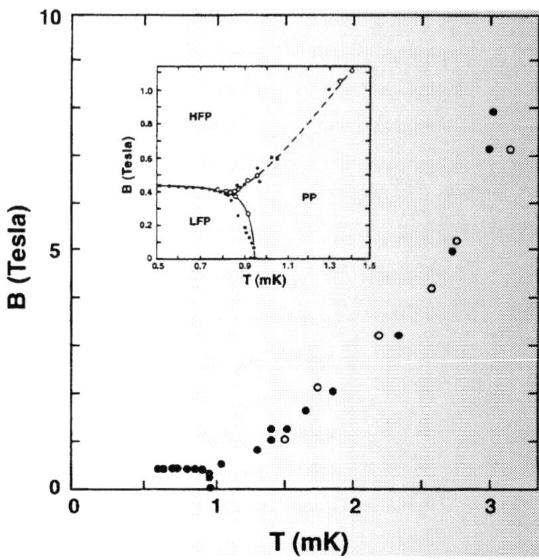

Fig. 5. The solid ^{3}He magnetic phase diagram at melting pressures as it was known in 1980 as Osheroff began his new studies of the spin ordered solid

Adams found for fields below about $0.4T$ that T_c decreased slowly as the field increased, while above $0.4TT_c$ increased rapidly as the field increased. He correctly concluded that there must be a second nuclear spin ordered solid phase at higher magnetic fields, whose magnetization was roughly two thirds of the maximum magnetization possible, when all the spins pointed in a single direction in space. By 1978 the known phase diagram for the ordered solid looked as it is shown in Fig. 5.

All this information had been extremely useful to the theorists, who postulated a spin Hamiltonian which involved two, three, and four atom exchange processes. Yet the values of the exchange parameters were not well known, and the magnetic sublattice structure of the ordered states were still quite a mystery. Indeed, one paper discussed thirteen possible ordered structures for the nuclear spins in the low field ordered phase [15], none of which turned out to be correct.

Except for the transition temperatures themselves, all existing information regarding the nature of the spin interactions in solid ^{3}He had been obtained well above the spin ordering temperature, where mean field theories can be used accurately, but where the effects due to spin–spin interactions are very small and hard to measure. I felt that it would be much more productive at this time to study the properties of the ordered phases directly. The solid was a very poor thermal conductor at such low temperatures, however, and the latent heat at the first order transition was relatively quite large. It was not believed that one could cool a bulk sample through T_c in any reasonable period of time.

I had read a research proposal by Bill Halperin several years before this time in which he proposed to grow ordered solid directly from the superfluid well below the solid T_c. Halperin had worried that to measure the temperature of the solid, he would need to grow it within the pores of platinum powder, and then use NMR to measure the platinum nuclear magnetization, which would vary as $1/T$. I didn't like this idea, and considered it likely that well below T_c the thermal conductivity in the solid would be due to antiferromagnetic spin waves. Since the solid was very pure, if the lattice structure was highly crystalline, the spin wave mean free path might be quite long, leading to a short thermal relaxation time within the spin system. Thus I decided to grow bulk samples in open regions of the cell, and measure their properties with NMR, which had been a very successful probe of the superfluid phases. I did use a powdered platinum NMR thermometer, but this was in contact only with the superfluid.

There was one problem with this strategy. In liquid ^{3}He, NMR proved useful in identifying the superfluid phases because of unexpectedly large frequency shifts resulting from the correlation between the atoms forming the Cooper pairs. While one might expect even larger frequency shifts in an antiferromagnet, that would not be true if the magnetic sublattice structure had a cubic symmetry, as the lattice itself did near the melting pressure. In this case the antiferromagnetic resonant frequency, which sets the scale for the frequency shifts, would vanish to first order, and any remaining shifts would be much smaller and very difficult to measure. While this concerned me, I have an unusually homogeneous superconducting magnet, which was designed to be particularly good at low fields where the frequency shifts, which vary roughly as $1/B$, would be largest.

I built a variable volume cell which functioned like a Pomeranchuk cell. The volume could be decreased by extending a metal bellows into the ^{3}He filled region, but it also contained a sintered silver heat exchanger which would allow me to pre-cool the liquid to about $0.4\,mK$, where the solid would be formed, using an adiabatic nuclear demagnetization cryostat. The solid growth process was initiated by raising the ^{3}He pressure slightly above the melting pressure, and discharging a small capacitor across a thin resistance wire, dumping $1erg$ into the liquid in about $1\,ms$. This region was surrounded by an NMR coil tuned to the frequency of the solid above T_c, about $1\,MHz$. The frequency was swept slowly over a $1\,kHz$ range as I searched for the ordered solid NMR signal.

At these temperatures, the B liquid NMR signal was unshifted and nearly temperature independent. There was in addition, generally a very small B liquid signal seen at a higher frequency, as a result of liquid crystal like textures in the superfluid. The all liquid NMR signal is shown in Fig. 6a. The solid signal should be considerably larger, and shifted by what I believed would be only a few Hz. At first I saw no evidence of any solid growing on my resistance wire at all. This was because I was not sweeping the NMR frequency far enough. I tried again, and again. Ultimately I saw a solid NMR peak, shifted

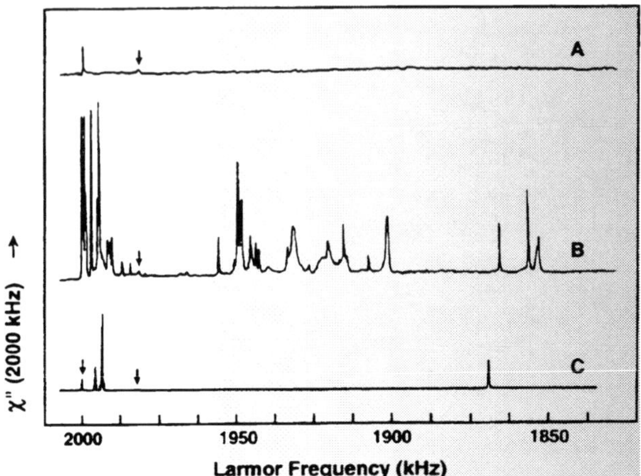

Fig. 6. (a) All B-liquid NMR spectrum at $0.4\,mK$. (b) NMR spectrum of a poly-crystalline solid ^{3}He sample grown at $0.4\,mK$ showing random orientations with respect to the static magnetic field. (c) NMR spectrum of a single crystal of solid ^{3}He at $0.4\,mK$, showing separate resonances from the three allowed magnetic domains

toward the high frequency limit of my sweep. I was elated, the frequency shift was nearly $1\,kHz$, vastly beyond my wildest dreams! I called Bob Richardson to give him the good news. Even though I have been away from Cornell for several years, I still had a close relationship with my ex-colleagues. However my call was premature. As I continued to study the NMR frequency over ever broader widths, I found additional resonance peaks shifted upward by as much as $150\,kHz$, with perhaps a total of 20 peaks visible in the NMR spectrum. A typical NMR spectrum of the solid grown in this way is shown in the Fig. 6b.

I felt that the spectrum was complex because I had grown a polycrystalline solid, with each crystallite oriented randomly with respect to the magnetic field. To make real progress, I needed to be able to grow a single crystal. I slowly increased the volume of my cell, melting most of my polycrystalline sample. Then, when I had only a tiny amount of solid left, I reversed the process. All subsequent solid grew on the single seed crystal remaining. I had succeeded. However what I say was perplexing. There was not just one but solid three NMR peaks visible, as shown in Fig. 6c. As I did this experiment over and over, I always found a minimum of three peaks, and the next smallest number was not four, but six. What was going on?

This beautiful puzzle was all worked out in about four months with the help of two Bell Labs theorists, Michael Cross and Daniel Fisher (now at Caltech and Harvard respectively) [16]. There is no way that I could have done it by myself. Their insights showed me what to measure, and my observations constrained their interpretations. Ultimately we obtained an NMR spectrum of a single crystal of ordered solid shown in Fig. 7. By measuring the limiting

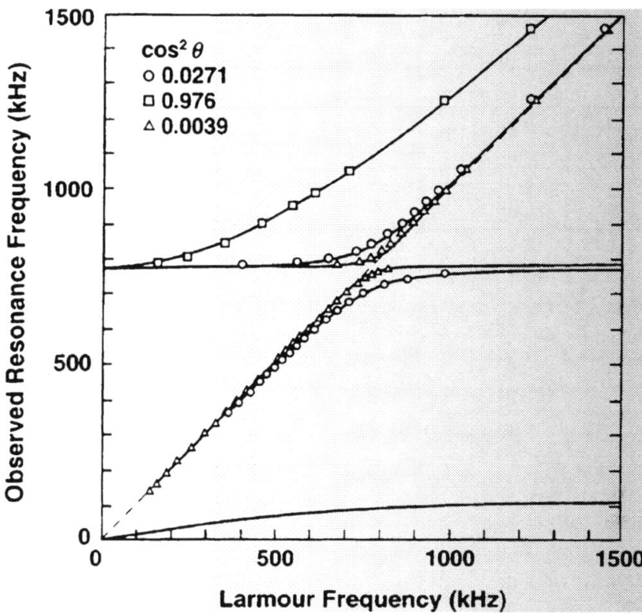

Fig. 7. Nearly complete NMR spectrum for a single crystal of spin ordered solid ^{3}He in low magnetic fields showing the three domain resonances above the Larmor frequency and two of the three domain resonances below the Larmor frequency

behavior of the NMR resonances at low and high magnetic fields, we were able to determine the symmetry of the sublattice structure, and to guess the precise structure, later confirmed by polarized neutron scattering [17]. The structure turned out to be rather interesting. The sublattices consist of planes of spins normal to any one of the principal lattice directions. The spins on these planes are oriented ferromagnetically, with the spin direction on successive planes alternating, two planes up and then two planes down. The three resonances described result from domains along the three principal directions in the body centered cubic lattice. This work was done while the movie Star Wars was still very popular, and I named this phase U2D2 in honor of the fearless robot R2D2.

The U2D2 phase of solid ^{3}He now serves as a useful model magnetic system. One can determine the orientation of the crystal in real space from the frequency shifts of the three domains, with a precision which rivals that possible with x-ray diffraction. The temperature dependence of the frequency shifts allows one to measure the temperature of the spin system with a resolution of about $100\,nK$, without having to attach a lead to the sample. Perhaps most importantly, solid ^{3}He exhibits a wonderful separation of energy scales, which makes observable phenomena much easier to interpret than is true for most ordered magnetic systems. The exchange energy, which results in the magnetic ordering, is four orders of magnitude smaller than the energy which localizes atoms to their lattice sites, and four orders of magnitude larger than the nuclear dipole-dipole energy which produces the NMR frequency shifts.

One may argue that this was not really a discovery, but simply incremental science. However I have described it here because virtually no one had believed there would be a large antiferromagnetic resonant frequency in the ordered solid. There is indeed not such a shift in the high field ordered phase. The usefulness of the resulting frequency shifts, both in terms of determining the sublattice structure and in terms of making the U2D2 phase a useful model magnetic system, were also not anticipated. I should add, however, that while the effort I was involved in at Bell Labs managed to do the most once we had observed the frequency shifts, a group headed by Dwight Adams at the University of Florida, using a Pomeranchuk refrigerator, had observed the shifted solid NMR at the same time, but quite independently [18]. We were able to get far more out of our data because of the higher flexibility of our cooling technique. Adams' group started growing solid ^{3}He at about $10\,mK$, and very little of the solid they formed was in thermal equilibrium. They also grew polycrystalline samples of solid, which could not be re-grown into a single crystal as we had done.

There are indeed lessons to be learned from this story: (1) Don't stay in any field too long. You get stale, and the impact of your work decreases. The work one is doing always appears (or should appear) new and exciting, but as time goes on, our perspective becomes increasingly myopic. (2) In deciding on a new research thrust, look for areas in which our understanding is rather incomplete. I suppose the less complete the better. (3) Try to study the system from a different perspective than had been used before. In our case, that meant looking directly at the properties of the ordered phase. (4) The combination of good experimentalists and good theorists makes for a far more effective research team than the two groups working separately. (5) Don't allow theoretical predictions to constrain your investigations. Even good theorists are far better at explaining observed physical behavior than they are at accurately predicting unobserved physical behavior.

I wish to point out that this 'discovery' depended crucially upon the work of many other people, and that the key idea, that of growing solid ^{3}He directly from the superfluid into the nuclear ordered solid, had been circulating through the low temperature community for years without being utilized, just as Pomeranchuk's suggestion had lay dormant for over a decade before it was tried. The trick is often finding some reason why an old idea should suddenly appear more likely to succeed. In my case it was the realization that the ordered solid would probably be a relatively good thermal conductor.

5 Weak Localization

In 1977 David Thouless had published a Physical Review Letter [19] proposing that if a sufficiently long and thin metallic wire were cooled well below $1\,K$, one would find the resistance to rise, rather than fall, and that ultimately it should rise exponentially with the resistance of the wire divided by the temperature. The physics responsible involved the interference of an electron wave function over different possible paths. This idea ultimately led to weak localization and to many phenomena seen in mesoscopic physics.

This article by Thouless had not escaped my notice, and when Joe Burton forced me to begin thinking about other areas of research, this one stood out in my mind as an exciting possibility. I was no expert at lithography, but low temperatures was my forte. One day I had lunch with Gerry Dolan, who had also joined Bell Laboratories from Cornell, and he, too, had been thinking about how to test the Thouless ideas. Gerry was an expert at fine line lithography. We quickly decided to team up.

At the time, I was beginning my work on solid ^{3}He, and I had only one cryostat in the lab. Dolan would make fine wires of Pd-Au and Cu-Au alloys, typically $30\,mm$ thick, on glass or sapphire substrates. These samples would have to wait until I was ready to cycle my cryostat up to room temperature and back down for other reasons before I would examine their low temperature properties. Dolan had developed clever techniques which allowed him to measure the resistance of his wires as he deposited them, thus insuring that he could get wires which met the Thouless criteria.

Once cold, we measured the I-V curves of the wires with a four wire technique as a function of temperature. Jerry's wires fell into two categories. Wires in the first category showed a very steep slope to the I-V curves for small applied voltages, but for higher voltages the slope was much shallower and independent of the applied voltage. This behavior suggested a tunnelling barrier somewhere along the length of the wire, and was not at all what we wanted. Wires in the second category showed essentially ohmic I-V curves with a slope which did not depend upon temperature. We found no other behavior after several iterations.

During this period of time, Phil Anderson, the preeminent condensed matter theorist at Bell Labs at this time, would come into my lab on almost a weekly basis to see what was going on. Phil had his name attached to one of the superfluid phases of liquid ^{3}He, and I listened to him often for advice. In 1977 he shared the Nobel Prize for his work on the electronic properties of disordered substances, and was an expert in the physics of what Dolan and I were trying to study. I would argue with Phil about why we didn't see what Thouless had predicted, but was never able to understand why things weren't working out.

Eventually, I became very frustrated. We had a set of samples in the cryostat whose resistance seemed quite ohmic, and independent of temperature. The current which we passed through our samples was generated by a slow function generator in series with a very large resistance at room temperature. The voltage was measured with a second set of sample leads using a high impedance low noise preamplifier which had a differential input, although I was only using one of the differential inputs, with the other grounded. I decided to subtract off as much of the voltage across our samples as I could, to see if there was anything left which was not strictly proportional to the current. To do this, I used a dummy sample at room temperature in series with another large resistor such that the voltage across the dummy sample resistor would be precisely proportion to the current, and nearly equal to that across the sample. This voltage was fed into the other input of the differential preamplifier. After I adjusted the new pair so that the two 'sample' voltages just cancelled, I swept the current and looked at one of my samples, which was at a very low temperature.

What I saw seemed strange. There was indeed a non-ohmic component to the resistance of our thin wire sample, but the shape of the I-V curve, shown in Fig. 8, was like nothing I had ever seen. I soon found that the shape of the I-V curve depended on the temperature of the wire, again in a way that I could not understand. I called Gerry down to my lab, and we both puzzled over the newly discovered behavior.

Soon Phil Anderson came into my lab, and asked what I was doing. I showed him the strange curves we had been obtaining, and he said without any hesitation "Why, that's a logarithm", and asked if he could borrow some of the data while he sat through a seminar. When he returned he had replotted our data, in a form which indeed made it look like a logarithmic dependence. He then pronounced that what we had discovered was 'weak localization' in two dimensions, as predicted by his unpublished theory. The final data are shown in Fig. 9.

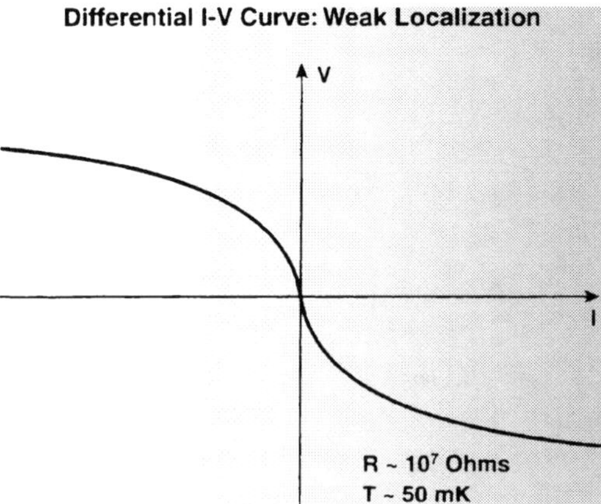

Fig. 8. I-V curve of a thin gold-palladium wire at about $50\,mK$, after subtracting off the ohmic portion

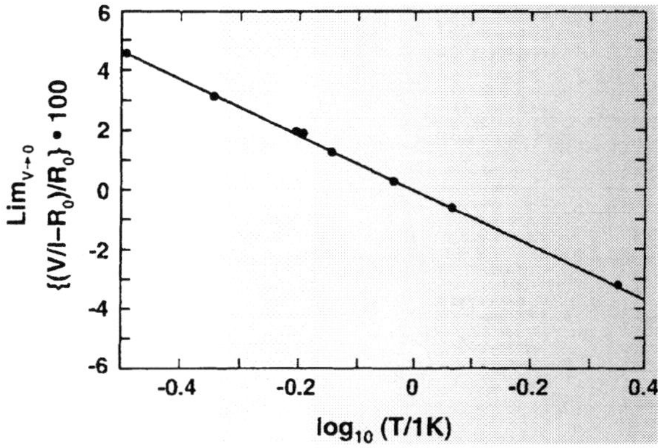

Fig. 9. Non-ohmic resistance of gold-palladium wire in the limit of small voltages vs. $log_{10}(T)$, showing behavior consistent with weak localization

Phil worked with Gerry and me to get as much out of our experiment as possible. As was usually the case, Phil was way ahead of me, and stayed there. Ultimately we published our experimental work and Phil's explanation back to back in Physical Review Letters [20]. This was the beginning of what was for a while a very active field in condensed matter physics, although it was my last work in the field. Soon the real experts at transport measurements dove in, and I decided to focus on the solid ^{3}He studies.

Gerry and I each gave invited talks at that year's March American Physical Society meeting. I talked on the solid ^{3}He work, and he talked on the weak localization work. I spoke to an audience of perhaps 250, and Gerry spoke to an overflow audience in excess of 700, in the same room the next morning. It was that day that I realized how little most physicists cared about quantum fluids and solids.

There are certainly lessons to this story: (1) Don't stay in any field too long. One gets stale, and the impact of their work decreases. The work we are doing always looks (or should look) new and exciting, but as time goes on, our perspective becomes increasingly myopic. I repeat this lesson here because it is the hardest one to learn! (2) Predictions by bright theorists are not always exactly correct, but experimentalists should always listen to them, because most have an excellent nose for where new physics is lurking. (3) Don't invent everything yourself. Build collaborations which utilize the strengths of different individuals. (4) Always think of how you can most clearly see what you intend to study. If I had not used a differential measurement to subtract off the ohmic signal, we would never have seen the unexpected one.

6 The Dipolar Gap In Amorphous Solids

I left Bell Labs in the fall of 1987 for Stanford University. This was an extremely difficult move for me to make, because I had been treated so well at Bell Laboratories. They believed it was in their best interest to give me the resources necessary to do the best work I could, and I had profited immensely from this wonderful support. In academia, universities pay your salary, but your research funds came from federal grants. I had always found it unpleasant to ask for resources, and have never been good at grantsmanship. To have a viable laboratory at Stanford it seemed clear, however, that I would have to carry out two separate research programs, since in condensed matter physics the grants were simply not large enough to support the four students I felt were necessary to provide a 'critical mass' of expertise. This meant I had to develop a new research program, clearly separate from my program in ^{3}He physics.

In 1989 high T_c superconductivity was definitely king in condensed matter physics, but I felt that the overall effort was too large, and that all the students trained in this field could not possibly find jobs in physics. Besides, I thought I should choose something which would rely on the same ultra-low temperature technology as the quantum fluids and solids program. The ground rules for choosing research projects funded by the federal government were different than those I operated under at Bell Labs. It would be hard for me to get two separate grants from the National Science Foundation, NSF, which was the only agency likely to support my quantum fluids and solids work. Again, I looked around for an area where I felt there was an incomplete understanding, but I wanted something which seemed like it might have applications at least elsewhere in physics. Applications were becoming important with federal agencies at this time.

Just how I decided what else to do is a long story for another time. Ultimately, however, I chose to study the low temperature properties of glasses. This was actually quite an old field. In 1970–71 Robert Pohl, at Cornell University, had discovered that all insulating glasses share a common set of anomalous low temperature properties [21], just one year before Lee, Richadson, and I were to discover superfluidity in liquid ^{3}He at the same institution. One of the anomalous properties of such materials is that the dielectric constant is temperature dependent down to very low temperatures, even below $2\,mK$. I proposed that such behavior could provide useful thermometry at very low temperatures, and could do so in a device with a very low heat capacity. Yet, at the same time, the technology was not well developed, and there were aspects of this low temperature glassy behavior which had never yet been understood. I proposed to develop that understanding and the technology necessary to make glass capacitance thermometry broadly useful. For reasons which I will not question, the US Department of Energy agreed to support my program, and I should add have been very supportive ever since.

After building a second major cryostat for studying physics at low temperatures, my Stanford graduate students and I began to look with high resolution at the dielectric properties of various amorphous samples. We soon began to realize that there was some aspect of our system over which we did not seem to have control. These were very slow measurements, and again I got frustrated. In a conversation with my students I proposed a very simple experiment to test our understanding. If we truly understood the origins of the dielectric behavior we were studying, the application of a large DC electric field across our capacitors should not change the AC dielectric constant we were measuring. My students tried the experiment, and what we saw was most puzzling. Upon the application of a large DC field, the capacitance shot up abruptly (by one part in 10^4). It then began a slow relaxation back, but this relaxation was not exponential in time, but logarithmic in time. This behavior is shown in Fig. 10a. We then decided to slowly sweep the DC electric field from $-7.10^6 V/m$ to $+7.10^6 V/m$. What we saw was again quite surprising. The AC dielectric constant had a minimum at $E = 0$, the field the sample had been sitting in for many days. If we stopped our sweep at some non-zero

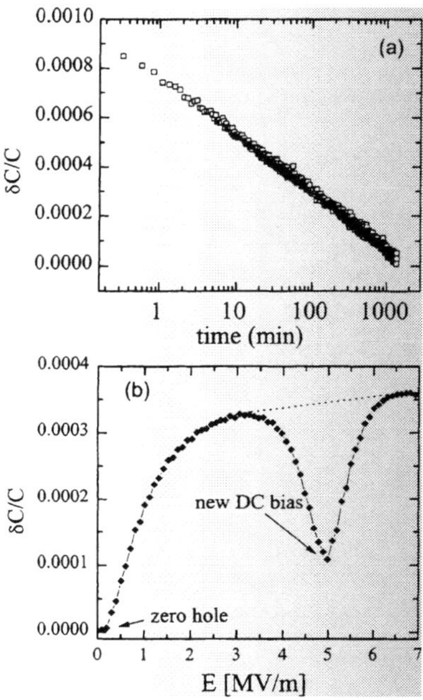

Fig. 10. (a) AC dielectric response of a glass sample following the application of a $3\,MV/m$ DC electric field, showing behavior logarithmic in the time since the application of the DC field. (b) AC dielectric response of a glass sample vs. DC electric field. The DC field was left at $5\,MV/m$ for one hour prior to this sweep. The growth of a new dielectric minimum (hole) is associated with the logarithmic relaxation seen in part (a)

electric field for a few hours, and then resumed our sweep, we found that a local minimum had formed in the dielectric constant at that field as well. This behavior is shown in Fig. 10b.

This second measurement had helped explain the first. The jump we saw when we quickly applied the DC electric field was equivalent to the change in dielectric constant we observed in the sweep experiment when we moved out of the minimum in C(E) which existed at E = 0, which we called the zero bias hole. The relaxation was then due to the formation of another minimum. What was going on here?

There is nothing like a good puzzle to keep physics interesting. This puzzle came to us in early 1992, just months before the biannual Phonons meeting, that year to be held at Cornell, hosted by Robert Pohl. On the way to this meeting, I stopped by Bell Laboratories and spoke to David Huse, who along with Daniel Fisher, had proposed one of two viable models of spin glasses. David suggested that the solution to my puzzle was physics very similar to what happened in a spin glass, where there is a hole in the distribution of magnetic moments vs. local magnetic field at zero local field. The argument is that if a spin resides in zero magnetic field, any perturbation, no matter how weak, will cause it to rotate. This will then change the magnetic field at the sites of the other spins, causing some of them to flip. This process will continue until a quasi-equilibrium state is found for which none of the magnetic moments resides in zero local magnetic field.

The situation is a bit more complex for a glassy insulator. For one thing, we have no magnetic moments. Instead, in 1972 it had been postulated that such glassy materials possess a broad distribution of what are called two level tunnelling states, TLS. These are molecules or a collection of molecules within the open structure of the glass which can reside in only two low lying energy states, as a function of their position or rotation. A tunnelling barrier exists between the two states. To recover the observed low temperature properties of glasses from such a model, one must postulate that the distribution of these TLS is flat, independent of the splitting between the two low lying energy states. This would be equivalent to an Ising spin glass with a random local magnetic field at the site of each spin. However, in glasses the interactions between TLS were always considered negligible compared to the splitting of the low lying energy states.

It took about two years to show that the rather simple idea by David Huse was more or less correct. However, to do so required a great deal of work, both experimental and theoretical. The reader is referred to a recent book which describes the nature and importance of interactions between thermally active defects in amorphous solids for a more complete picture of this issue [22].

This is an exploration still in progress. There are, none the less, lessons to this story as well: (1) Again, seek out a system for which no complete understanding appears to exist. (2) In experimental studies, don't just make the measurements you have intended. It takes very little extra time to devise new tests, which may show evidence for new behavior far more clearly than the measurements you had intended to make for other reasons. Experimentalists with blinders on may work very efficiently at what they have planned, but will miss the gems which lay hidden just beyond their notice.

7 Conclusions

I have tried to convince you that there are research strategies which can greatly increase the probability of 'serendipitous' discovery, and that they need not involve wild risk-taking. For most of these strategies, however, there remains the question of degree. For example, the hardest thing for an experimentalist to decide is when to leave a study and move on to something new. Being the worlds expert at something may insure an ability to do good incremental research, but may make major breakthroughs less likely.

One interesting question which often plagues graduate students is the following: How much must he or she know about a subject in order to contribute to mankind's overall knowledge of that subject? If one knows too little, they are likely to miss the subtle indications of something new, or dismiss such evidence as an experimental artifact. If one knows too much, however, one's mind may become constrained by current wisdom on the subject. My own policy (and nothing more than that) has always been that one should *understand the subject well enough to acquire a good physical intuition about how it should behave. However, when it comes to one's instrumentation, it is necessary to understand how it functions and what it actually measures in minute detail. One can never understand one's equipment too well. Only with a firm understanding of the meaning of one's measurements, and an abiding confidence in that understanding, is one likely to recognize the indications of new physics and spend the time to follow up on those indications.

Let me close by addressing the question: 'Does our current system of research support in this country encourage or discourage the behavior most likely to result in discovery?' I think the answer is a complex one. In general, I believe research funding agencies recognize the value of new technologies, and tend to support its development. At the same time, the requirement for frequent progress reports and grant renewal proposals makes it hard for people to change the subject of their research, or to pursue anything which is not likely to produce results in a two year time frame. If an investigator fears that his grant will not be renewed unless he produces a slug of new publications, he will focus on those things he knows how to do and feels fairly confident will yield results in a timely manner. Worse still, this pressure simply does not produce the atmosphere in the lab most conducive to the training of bright graduate students. Students need to be given guidance, but as much freedom as they can handle, including the freedom to make occasional mistakes. Too often graduate students are being treated like technicians, not scientists. I hesitate to think of where I might be now if I had not been given a great deal of freedom by my thesis advisor, David Lee. In addition, most individual investigator research grants today are rather lean, and do not provide the flexibility to allow investigators to pursue both the physics which they have outlined in their proposal (typically fairly safe incremental research) and physics which lies beyond the heart of their research programs. The optimization of research funding involves many complex issues, both political as well as scientific in nature. The scientific community itself, however, needs to recognize these issues, and do what it can to facilitate a research climate conducive to discovery and to the most effective training of our next generation of scientists. If these two things are done, I am confident that exciting new physics will continue to be found.

References

1. D. Goodstein and J. Goodstein, *"Richard Feynman and the History of Superconductivity"* in *History of Original Ideas and Basic Discoveries in Particle Physics*, ed. H.B. Newman and T. Ypsilantis, Plenum, N.Y. (1996), pp. 773–779.
2. J. Bardeen, L.N. Cooper, and J.R. Schrieffer, "Theory of Superconductivity", Phys. Rev. **108**, 1175–1204 (1957).
3. R.W. Wilson, *The Cosmic Microwave Background Radiation* Les Prix Nobel 1978, eds. Siegbahn, K., et al., Almqvist & Wiksell, Stockholm, (1979) pp. 113–133.
4. G. Gamow, "The Evolution of the Universe", *Nature* **162**, 680–682 (1948).
5. D. Wilkinson, private communication.
6. I. Pomeranchuk, "On the theory of liquid ^{3}He", *Zh. Eksperim. i Theor. Fiz.* **20**, 919–926 (1950).
7. Y.D. Anufriyev, "Use of the Pomeranchuk effect to obtain infralow temperature", *JETP Lett.* **1**, 155–157 (1965).
8. W.E. Keller, *Helium-3 and Helium-4*, Plenum, N.Y. (1969).
9. R.T. Johnson, R.E. Rapp and J.C. Wheatley, "Effect of a magnetic field on the melting curve of ^{3}He", *J. Low Temp. Phys.* **6**, 445–453 (1971).
10. D.D. Osheroff, *Superfluidity in ^{3}He: Discovery and Understanding*, Les Prix Nobel 1996, eds. T. Frngsmyr and Brigitta Lundeberg, Norstedts Tryckeri AB, Stockholm (1997) pp. 103–133. Also: *Rev. Mod. Phys.* **69**, 667–681 (1997), and available from World Scientific Publishers (Singapore) on CD ROM.
11. D.D. Osheroff, "Nuclear magnetic order in Solid ^{3}He", *J. Low Temp. Phys.* **87**, 297–342 (1992).
12. N. Bernardes and H. Primakoff, "Theory of Solid ^{3}He", *Phys. Rev. Lett.* **2**, 290–292 (1960).
13. J.R. Sites, D.D. Osheroff, R.C. Richardson and D.M. Lee, "Nuclear magnetic susceptibility of solid ^{3}He cooled by compression from the liquid phase", *Phys. Rev. Letts.* **23**, 836–839 (1969).
14. W.P. Halperin, C.N. Archie, F.B. Rasmussen, R.A. Buhrman and R.C. Richardson, "Observation of nuclear magnetic order in solid ^{3}He", *Phys. Rev. Lett.* **32**, 927–930 (1974).
15. M. Roger, J.M. Delrieu and A. Landesman, "Nuclear spin ordering with four spin exchange in solid bcc ^{3}He", *Phys. Lett.* A **62**, 449–452 (1977).

16. D.D. Osheroff, M.C. Cross and D.S. Fisher, "Nuclear Antiferromagnetic Resonance in Solid ^{3}He", *Phys. Rev. Lett.* **44**, 792–795 (1980).

17. A. Benoit, J. Bossy, J. Flouquet and J. Schweizer, Magnetic diffraction in solid ^{3}He", *J. de Physique Letters*, **46**, L923–L927 (1985).

18. E.D. Adams, E.A. Schubert, G.E. Haas, and D.M. Bakalyar, "NMR in magnetically ordered solid ^{3}He", *Phys. Rev. Lett.* **44**, 789–792 (1980).

19. D. Thouless, "Maximum metallic resistance in thin wires", *Phys. Rev. Lett.* **39**, 1167–1170 (1977).

20. P.W. Anderson, E. Abrahams, and T.V. Ramakrishnan, "Possible explanation of nonlinear conductivity in thin-film metal wires", *Phys. Rev. Lett.* **43**, 717–720 (1979) and G.R. Dolan and D.D. Osheroff, "Non-metallic conduction in thin metal films at low temperatures", *Phys. Rev. Lett.* **43**, 721–724 (1979).

21. R.C. Zeller and R.O. Pohl, "Thermal conductivity and specific heat of non-crystalline solids", *Phys. Rev.* B **4**, 2029–2041 (1971).

22. *Tunnelling Systems in Solids*, ed. P. Esquinazi, Springer (1998).

Symmetry in the Micro World –
A Conversation with Nobel Laureate
Eugene Wigner

B.G. Sidharth

B.M. Birla Science Centre, Adarsh Nagar, Hyderabad, India

Prof. Eugene Wigner was considered to be a superstar of physics along with his brother-in-law P.A.M. Dirac and the legendary R.P. Feynman. Eugene Paul Wigner's decades of yeomen service to the cause of comprehending nature in its barest, most fundamental aspect climaxed in the 1963 Nobel Prize for physics.

He was born in Budapest, Hungary in 1902 and went to earn his PhD at the Technische Hochschule in Berlin. Wigner who had already rubbed shoulders with legends like Dirac, Jordan, Niels Bohr, Wernher Heisenberg and Albert Einstein, became an American citizen in 1937. From 1938 he was Thomas D Jones Professor of mathematical physics at Princeton University, till his retirement in 1971. During these distinguished decades, Prof. Wigner participated in the celebrated World War II Manhattan Project at the University of Chicago. Thereafter he directed research and development at the Clinton laboratories. Prof. Wigner has been honoured with an endless string of awe-inspiring awards. From the US medal for merit in 1946 to the National Medal of Science in 1969. Not to mention honorary doctorates, and memberships of prestigious scientific bodies. He died in 1995.

I met him when he was 83, at a Conference on Group Theory. Not only was he very much alive and kicking, he was still a humble student of science, very gentle and very self effacing. Renowned scientists would spring to their feet on seeing him, in awe and respect. At one of the sessions in the Conference, which Prof. Wigner was attentively following, he got up and asked how the author had got a particular result. The author replied that this was because of the extra dimensions of spacetime. "What are these extra dimensions" Prof. Wigner asked. The author tried to explain at length how extra dimensions would come in. Prof. Wigner heard attentively for a while and said abruptly, "I do not understand all this", and sat down.

Despite the reputation as being a political hawk, Prof. Wigner was actually a very humane and affectionate person and a doting grandfather. He seldom got into a photograph without tacking a photograph of his grandchildren on to his coat.

B.G. Sidharth (ed.), *A Century of Ideas*, 205–207.
© Springer Science+Business Media B.V. 2008

Another interesting thing about Prof. Wigner was the fact that in spite of his long sojourn in the United States, his English was not as impeccable as his physics – it still had a quaint continental flavor.

We had a long meeting on the sidelines of the Conference and I reproduce a summary of the conversation.

"Starting from the beginning, how did you enter physics – and then get so deeply involved in it?"

"I had my degree in chemical engineering – jobs for physicists were few. So my father persuaded me to study chemical engineering. But my interest was in physics. I attended physics colloquia. But my doctorate degree was in physical chemistry, it was on the rate of reactions. There was more physics than chemistry in it. I got my PhD from the Institute of Technology at Scharlottenburg, near Berlin. Then I got a job in Hungary, in a leather tannery and worked for a little less than two years. But I was subscribing articles to the German journal, *Zeitschrift fur Physik.*

"One day I read an article of (Niels) Bohr and (Von P) Jordan which conveyed a new idea that was essentially the basis of quantum mechanics. It was based on Heisenberg's article which I had not read. At the same time I got an offer to become assistant to a (physics) professor at the same institute – and accepted it. That is how I officially got into physics. Before that I doubted if man was bright (capable) enough to understand microscopic phenomena. Two years earlier in a conversation, my brother-in-law Dr. Dirac, told me that before the article (I have referred to) came out he also had the same doubt. But that article convinced me that it is possible (for man to understand microscopic phenomena) – in 1927 when I was about 25. But later, Schrodinger's equation was more wonderful."

"Did any scientist inspire you?"

"I was assistant to Richard Becker. He taught me how to explain something. He was a good teacher, well versed in many areas of physics."

"What exactly put you on to the topic in which you are an acknowledged master – namely the symmetry that is buried deep in the principles that govern the universe?"

"I read another of Heisenberg's article in which he connected the symmetry of wave functions at rest to interchange particles. He worked it out for two particles and I worked it out for four particles. Becker encouraged me to publish it – and then Von Neumann suggested that I read representation theory."

"For which specific work were you awarded the Nobel Prize?"

"The Nobel Prize was for the discovery of symmetry principles and application to atomic and nuclear spectra – but I am not sure."

"Do you think that was your best work?"

"I don't know. Einstein got the Nobel Prize – and it was not (for) his best work. I don't know."

"Exactly when and how did you get the news of your winning the Nobel Prize?"

"I was having breakfast at Oakridge, Tennessee and someone came – I don't remember who – and told me, "Congrats" and I asked him, "Thank you, what for?" He said, "You got the Nobel Prize". I asked "How did you know?" (He replied) "I heard it on the radio.""

"Soon after, I returned to Princeton and stopped at Washington on the way. A reporter asked me what the Nobel Prize was for. "I wish I knew", I replied. It was one of the quotes of the day, the next day, in the New York Times.""

"Do you think we can understand the universe better? Or that we are understanding it better?"

"We are understanding it better. We can now understand a larger part of nature – but understand is not a good word. It is a better and more compact description of some of the inanimate aspects of nature. But not yet all the aspects. There are some problems for example in high energy phenomena, we cannot yet describe. And we cannot yet describe life."

"By life you mean consciousness"

"Yes, but I don't know if we can describe it. It's wonderful we can describe so many phenomena (at all). We must strive to do it."

"Prof. Wigner, do you think we can, in principle, understand all of inanimate nature?"

"I hope not – we will understand it better and better, but I hope it won't be fully, because if we knew everything there would be no interest left to do research."

"Do you think there is God?"

"I don't know."

"Who put forward the greatest ideas – at least in the world of physics?"

"Newton, Galileo, Maxwell, Planck, Heisenberg, Bohr, Jordan, Schrodinger and Dirac. Einstein contributed enormously also."

"Talking about Einstein, what is your opinion about his never being able to reconcile himself to quantum mechanics?"

"Einstein was mistaken. His argument has a philosophical implication which is incorrect. I liked Einstein and admired him, but in this respect he was conservative."

"What are your apprehensions about the future? How would you like the world to be?"

"I am afraid that one government conquers the earth and man becomes an inhabitant of an ant heap. It is good if there are several nations and man can move from one country to another according to the United Nations Treaty. And every government should look to the happiness of the people."

"One last question, Professor. Have you ever been to India?"

(Clapping his thighs and nodding almost sadly) "It was a great mistake. A great mistake – but I have never been to India.""

Made in the USA
Monee, IL
03 February 2021

2e60d07b-ee5f-438d-bc9f-78bc48ac8b26R01